How Species Interact

How Species Interact

Altering the Standard View on Trophic Ecology

Roger Arditi

and

Lev R. Ginzburg

OXFORD
UNIVERSITY PRESS

OXFORD
UNIVERSITY PRESS

Oxford University Press, Inc., publishes works that further
Oxford University's objective of excellence
in research, scholarship, and education.

Oxford   New York
Auckland   Cape Town   Dar es Salaam   Hong Kong   Karachi
Kuala Lumpur   Madrid   Melbourne   Mexico City   Nairobi
New Delhi   Shanghai   Taipei   Toronto

With offices in
Argentina   Austria   Brazil   Chile   Czech Republic   France   Greece
Guatemala   Hungary   Italy   Japan   Poland   Portugal   Singapore
South Korea   Switzerland   Thailand   Turkey   Ukraine   Vietnam

Copyright © 2012 by Oxford University Press

Artwork by Amy Dunham and Ksenia Golubkov © 2011 Applied Biomathematics
Published by Oxford University Press, Inc.
198 Madison Avenue, New York, New York 10016

www.oup.com

Oxford is a registered trademark of Oxford University Press

All rights reserved. No part of this publication may be reproduced,
stored in a retrieval system, or transmitted, in any form or by any means,
electronic, mechanical, photocopying, recording, or otherwise,
without the prior permission of Oxford University Press.

Library of Congress Cataloging-in-Publication Data
Arditi, Roger.
How species interact : altering the standard view on trophic ecology / Roger Arditi and Lev Ginzburg.
p. cm.
Includes bibliographical references and index.
ISBN 978-0-19-991383-1 (hardcover : alk. paper) 1. Predation (Biology)—Mathematical models.
2. Food chains (Ecology)—Mathematical models. I. Ginzburg, Lev R. II. Title.
QL758.A73   2012
591.5'3—dc23   2011034905

1 3 5 7 9 8 6 4 2

Printed in the United States of America
on acid-free paper

Ce qui est simple est toujours faux. Ce qui ne l'est pas est inutilisable.
Paul Valéry, in *Mauvaises Pensées et Autres* (1942)

[*What is simple is always wrong. What is not is useless.*
Paul Valéry, in *Bad Thoughts and Others* (1942)*]*

We must learn from the mathematician to eliminate and to discard; to keep the type in mind and leave the single case, with all its accidents, alone; and to find in this sacrifice of what matters little and conservation of what matters much one of the peculiar excellences of the method of mathematics.
D'Arcy W. Thompson, in *On Growth and Form* (1917)

CONTENTS

Acknowledgments   *xi*
Acronyms and Symbols   *xiii*

Introduction   3

1. Alternative Theories of Trophic Interaction   8
     1.1 Monod versus Contois: Resource-Dependent and Ratio-Dependent Bacteria   9
     1.2 The Standard Predator-Prey Model of Ecology   14
     1.3 The Arditi-Ginzburg Ratio-Dependent Model   17
     1.4 Donor Control and Ratio Dependence   22
     1.5 Predator-Dependent Models   24
     1.6 What Happens at Low Density? The Gradual Interference Hypothesis   26
     1.7 Biomass Conversion   29

2. Direct Measurements of the Functional Response   33
     2.1 Insect Predators and Parasitoids, Snails, Fish, and Others: Laboratory Measurements   35
         2.1.1 Manipulating the Consumer Density Alone   35
         2.1.2 Measuring Interference in the Presence of a Saturating Functional Response   37
         2.1.3 The Arditi-Akçakaya Predator-Dependent Model   40
         2.1.4 Application to Literature Data   41
         2.1.5 Does Interference Increase Gradually?   45
     2.2 Wasps and Chrysomelids: A Field Experiment   45
     2.3 Wolves and Moose: Field Observations   48
         2.3.1 Wolf Social Structure and Spatial Scales   49
         2.3.2 Model Fitting and Model Selection Methods   50
         2.3.3 The Wolf-Moose Functional Response Is Ratio Dependent   51

2.4 Additional Direct Tests of Ratio Dependence   55
    2.4.1 Bark Beetles   56
    2.4.2 Shrimp   57
    2.4.3 Egg Parasitoids   57
    2.4.4 Benthic Flatworms   59
2.5 Identifying the Functional Response in Time Series   60
2.6 Concluding Summary   61

3. Indirect Evidence: Food Chain Equilibria   62
    3.1 Cascading Responses to Harvesting at the Top of the Food Chain   63
    3.2 Enrichment Response When the Number of Trophic Levels is Fixed   66
    3.3 Enrichment Response When the Number of Trophic Levels Increases with Enrichment   71
    3.4 Why the World is Green   76
    3.5 The Paradox of Enrichment   77
    3.6 Donor Control and Stability of Food Webs   80

4. How Gradual Interference and Ratio Dependence Emerge   83
    4.1 Experimental Evidence of the Role of Predator Clustering on the Functional Response   84
        4.1.1 An Aquatic Microcosm Experiment   85
        4.1.2 Predator Aggregations Lead to Ratio Dependence   86
    4.2 Refuges and Donor Control   89
        4.2.1 A Simple Exploratory Theoretical Model   90
        4.2.2 From Donor Control to Ratio Dependence   92
    4.3 The Role of Directed Movements in the Formation of Population Spatial Structures   93
        4.3.1 Self-Organization Due to Accelerated Movement   94
        4.3.2 Spatially Structured Predator-Prey Systems   96
        4.3.3 How Ratio Dependence Emerges From Directed Movement   101
    4.4 Ratio Dependence and Biological Control   102
        4.4.1 The Biological Control Paradox   102
        4.4.2 Trophotaxis and Biological Control   103
    4.5 Emergence of Gradual Interference: An Individual-Based Approach   108
        4.5.1 A Qualitative Model Based on Predator Home Ranges   108
        4.5.2 An Individual-Based Model Based on Trophotaxis   111

5. The Ratio Dependence Controversy   115
    5.1 Evidence of Ratio Dependence is Often Concealed in the Literature   116
    5.2 The Paradox of Enrichment and the Cascading Enrichment Response: Is There Any Evidence That They Exist?   117
    5.3 The Fallacy of Instantism   120
    5.4 How the Ratio-Dependent Model Serves the Debate on the Causes of Cyclicity   124
    5.5 Mechanistic versus Phenomenological Theories   126
    5.6 "The Truth is Always in the Middle": How Much Truth is in This Statement?   127

6. It Must Be Beautiful   129
    6.1 Scaling Invariance and Symmetries   130
    6.2 Kolmogorov's Insight   135
    6.3 Akçakaya's Ratio-Dependent Model for Lynx-Hare Cycling   137
    6.4 The Limit Myth   139

Appendixes   143
    3.A Food Chain Responses to Increased Primary Production   143
        3.A.1 Prey-Dependent Four-Level Food Chain   143
        3.A.2 Ratio-Dependent Three-Level Food Chain   145
    3.B Cascading Response in the Ratio-Dependent Model   146
    6.A How a Revised Ecology Textbook Could Look   148

References   151
Index   163

ACKNOWLEDGMENTS

We are grateful to our coauthors in a number of articles whose results are reported in this book, and particularly to Reşit Akçakaya, Chris Jensen, Christian Jost, and Yuri Tyutyunov. The following persons made a number of insightful comments after reading all or part of the manuscript: Reşit Akçakaya, Don DeAngelis, Nick Friedenberg, Bob Holt, Nicolas Loeuille, Stuart Pimm, Thierry Spataro, Yuri Tyutyunov, and two anonymous reviewers. Reşit Akçakaya also made available the unpublished results used in appendix 3.B. LRG's students Spencer Koury, Emily Rollinson, James Soda, and Omar Warsi suggested a number of changes to make the text more accessible and pedagogic. Amy Dunham drew the portraits of Volterra, Lotka, and Gause, and Ksenia Golubkov drew those of Monod, Contois, and Kolmogorov. John Vucetich gave permission to print on the cover his photograph of a pack of wolves attacking a moose in Isle Royale National Park. Monique Avnaim helped assemble the list of references and Clarisse Coquemont searched libraries for older documents. Lara Borrell and Haeju Kim Lundquist were the essential link in the daily communication between the two authors. RA thanks CNECA2 (French Ministry of Agriculture) for granting a full one-year sabbatical leave to work on this book.

RA dedicates this book to the memory of his wife, Jacy Arditi-Alazraki (1948–2006). Jacy used to say that life is short and that one must do what one has to do, say what one has to say, despite obstacles. After the shock of her sudden death, RA found in Jacy's advice the energy to complete first her own unfinished book[1], then to write the present book, a project that had lain dormant for years.

---

1. *Un certain savoir sur la psychose: Virginia Woolf, Herman Melville, Vincent van Gogh*, Paris: L'Harmattan, 2009.

ACRONYMS AND SYMBOLS

Acronyms

LV	Lotka-Volterra dynamic model or Lotka-Volterra functional response
Ho	Functional response model of Holling (1959b)
HV	Functional response model of Hassell and Varley (1969)
AA	Functional response model of Arditi and Akçakaya (1990)
DAB	Functional response model of DeAngelis et al. (1975) and Beddington (1975)
AG	Functional response model of Arditi and Ginzburg (1989)
AG-DC	Donor control functional response model (as defined by Arditi and Ginzburg 1989)

Symbols

The list includes only symbols that are used repeatedly in different chapters.

N	Abundance of prey (numbers, biomass, or density as appropriate)
P	Abundance of predators (numbers, biomass, or density as appropriate)
t	Time
r	Prey intrinsic growth rate
K	Prey carrying capacity
$g(\cdot)$	Functional response (as a function of unspecified variables)
a	Searching efficiency (sometimes called attack rate)
α	Prey availability (in AG, AG-DC, and AA functional responses)
h	Handling time
m	Mutual interference coefficient (in HV and AA models)
e	Conversion efficiency of killed prey into predators
q	Predator intrinsic death rate

How Species Interact

Introduction

The standard theory of predator-prey interactions taught in the common textbooks has changed little since the time of Lotka and Volterra in the 1920s. The main improvement was the introduction of Holling's generalization of functional responses to nonlinear forms, together with the prey carrying capacity, which led to the paradox of enrichment and the limit cycles of the more complex Rosenzweig-MacArthur model.

The authors of this book attempted to suggest substantial changes in the 1970s, publishing independently first in Russian and in French. The implications of these suggestions went mostly unnoticed until our first common article on the ratio-dependent interaction (Arditi and Ginzburg 1989), which started a heated controversy. We actually wrote the article in 1987 but it took two years and three journal submissions to see it through publication. Today, this article is highly cited, and its citation rate has been increasing continuously for over 20 years.

Our view was not without antecedent. In our 1989 article and in later ones, we brought attention to important articles of the 1970s by J. R. Beddington, D. L. DeAngelis, M. P. Hassell, and a few others that also moved away from the Lotka-Volterra assumption in a direction similar to ours. These authors had proposed models accounting for the influence of predator density (so-called interference). However, while these other authors generally presented interference as a somewhat annoying complication, we view it as a fundamental feature governing predator-prey interactions. A forerunning attempt to amend the Lotka-Volterra model can also be found in the work of Leslie (1948). In his model, the predator equation was improved in such way that the predators' specific growth rate became ratio dependent. However, the prey equation was not changed, leading to a logical inconsistency (see section 1.7). In a sense, our work can also be interpreted as a continuation of Leslie's, generalizing the same ratio-dependent view to the whole dynamic model. Even earlier, Kolmogorov (1936) hinted to this possibility (see section 6.2).

Left: Vito Volterra (b. Ancona 1860; d. Rome 1940). Born into a very poor family, he became a mathematician with a high reputation in the theory of functionals and integro-differential equations. In 1905, he was named as a senator by the king of Italy. At the age of 65, he developed an interest in modeling the dynamics of interacting populations, stimulated by his son-in-law, a biologist. At about the same time, he refused to take an oath of loyalty to Mussolini and was forced to resign his university post and all academic positions. He collected his major contributions to ecological theory in the book *Leçons sur la théorie mathématique de la lutte pour la vie* (1931). Right: Alfred James Lotka (b. Lemberg [today Lviv in Ukraine] 1880; d. New York 1949). Born to an American missionary couple, he received an international education (England, Germany, United States). His main interest was demography. While visiting Johns Hopkins University, he published the book *Elements of Physical Biology* (1925). He then spent most of his career as a statistician for Metropolitan Life Insurance in New York. Volterra and Lotka independently developed the predator-prey model known as the Lotka-Volterra equations, for which they are famous today. For both, this contribution was secondary to their major interests. See Kingsland (1985) and Wikipedia for additional historical information. Drawings by Amy Dunham.

In retrospect, we note that, every five years, we have reviewed the state of the debate, showing how some misunderstandings were gradually resolved and how a number of points that we had made were more easily accepted (Akçakaya et al. 1995; Abrams and Ginzburg 2000; Jensen and Ginzburg 2005). While the main ideas were first presented more than 20 years ago, we continuously brought novel developments. This book is the opportunity to bring together all our considerations, reorganized as a consistent demonstration, with a sequence of logical arguments rather than following a chronological order.

Ratio-dependent predation is now covered in major textbooks as an alternative to the standard prey-dependent view (e.g., Krebs 2009, 194; Gotelli 2008, 148; Begon et al. 2006, 304–305). One of this book's messages is that the two simple extreme theories, prey dependence and ratio dependence, are not the only alternatives: they are the ends of a spectrum. There are ecological domains in which one view works better than the other, with an intermediate view being also a possible case. Hanski (1991, 142) described the meaning of our proposed change in the basic model of predation with the following words:

> During the past few decades, theoretical ecology has raised a formidable tower of connected models: a platform from which ecologists might comfortably observe the tangled bank and its inhabitants below, and make some sense of their doings. The platform of models has been constructed by adding all sorts of conceivable, and often quite a few inconceivable, modifications to previous, simpler and in some sense more fundamental assumptions. One problem is that some of these fundamental assumptions are just that—assumptions; another tower of equal height and interest to ecologists could be raised using different assumptions. A more serious problem is that the views from the two towers might show profound discrepancies, even if the same bank were observed.

Our years of work spent on the subject have led us to the conclusion that, although prey dependence might conceivably be obtained in laboratory settings, the common case occurring in nature lies close to the ratio-dependent end. We believe that the latter, instead of the prey-dependent end, can be viewed as the "null model of predation." Hence the subtitle we have chosen for this book.

We start the first chapter with a clear example from microbiology, the Monod versus Contois uptake expressions describing the growth rate of bacteria as a function of sugar concentration. The Monod model assumes that bacterial growth is determined by the absolute concentration of sugar while the Contois model assumes that bacterial growth responds to per capita sugar concentration. We show how both are correct for different ranges of bacterial densities. We then move to ecology with the standard model and the ratio-dependent alternative. Both can be seen as limits of more general models. Particularly, we propose the gradual interference model, a specific form of predator-dependent functional response that is approximately prey dependent (as in the standard theory) at low consumer abundances and approximately ratio dependent at high abundances. This can formalize the somewhat vacuous view that "the truth is in the middle." We also emphasize one logical requirement that predator-prey models must obey: the equations must express the fact that predator biomass is a conversion of prey biomass.

Since predictions of the two extreme views are quite different, both for equilibrial and for dynamic properties, at the population and at the community levels, we summarize three kinds of evidence that support our proposed predation theory. In chapter 2, it is all direct evidence, studies in which the number of prey eaten was measured either in the laboratory or in the field, on taxa ranging from insects to big mammals. The identification of functional responses raises specific mathematical and statistical difficulties. We show that careful attention to some overlooked technical problems requires reexamination of several previous analyses, leading to a reassessment that is much in favor of ratio dependence. Chapter 3 presents indirect evidence, mostly from lakes and marine systems, regarding the responses of food chains to enrichment. The two extreme views make contrasting predictions as to the equilibrial responses of the successive trophic levels with respect to enrichment at the bottom. We review and discuss a number of empirical observations. On the theoretical side, we discuss the "trophic cascade" paradigm. Food webs based on donor control (a special case of ratio dependence) are shown to be much more stable than those based on Lotka-Volterra interactions. This can explain the persistence of complex food webs in nature.

Chapter 4 is devoted to mechanistic theoretical approaches explaining how ratio dependence emerges at the global scale from various behavioral models of species interaction, even when assuming prey dependence at the local scale. Refuges and spatial heterogeneity as well as forms of temporal and biological heterogeneities can all lead to a ratio-dependent functional response. Individual-based modeling shows that the latter also emerges when the movements of predators obey some realistic directional rules.

Chapter 5 reviews the controversy surrounding the topic of this book and its current resolution. Finally, chapter 6 concludes the presentation by explaining the role of the ratio-dependent property as an invariance in ecology. It is equivalent to having scale invariance with respect to Malthusian growth of two interacting populations. Like all invariances, it is fundamentally imprecise but simple and useful. The book *Ecological Orbits* (Ginzburg and Colyvan 2004) had already discussed this invariance, which is necessary for consistency of predation theory with exponential growth of single species. The main emphasis of that earlier book—inertial growth—is best understood after one recognizes the need for such invariance. In a sense, the two books, *How Species Interact* and *Ecological Orbits* should be read in the reverse order of publication, as the inertial growth idea can be seen as the second installment of a related story.

In sum, this book presents three lines of argument that should answer the interests of scientists with different inclinations: chapters 2 and 3 discuss the evidence of data, either experimental or observational; chapter 4 discusses the underlying mechanisms, particularly with individual-based modeling; and chapter 6 discusses theoretical questions of symmetry and logical consistency. Our investigations deal with the foundations of ecology and can have profound consequences for our understanding of food chain structure and dynamics. Since the issues of food webs (mostly equilibrial properties but sometimes dynamic ones too) are quickly moving from purely academic to practical (eco-manipulation, biological control), we expect that this book will be useful to ecologists inside and outside academia.

CHAPTER 1

Alternative Theories of Trophic Interaction

Studies of food chains are on the edge of two domains of ecology: population and community ecology. The properties of food chains are determined by the nature of their basic link, the interaction of two species, a consumer and its resource, a predator and its prey.[1] The study of this basic link of the chain is part of population ecology while the more complex food webs belong to community ecology. This is one of the main reasons why understanding the dynamics of predation is important for many ecologists working at different scales.

An even stronger reason why predation is so important is that it is the most fundamental ecological interaction: no organism can live, grow, and reproduce without consuming resources (Murdoch et al. 2003, 1–2). Some ecologists even consider that the other commonly classified ecological interactions (competition, mutualism, commensalism, amensalism, etc.) can all be formalized as the influences of a third agent on the more fundamental interaction between a focal population and its limiting resource.

As we explained in the introduction, this book focuses on the alternative ways to describe elemental trophic interactions mathematically. In 1989, we published an article in which we classified the various models as *prey dependent*, *predator dependent*, and *ratio dependent* (Arditi and Ginzburg 1989). A controversy ensued in the literature but, as with most scientific issues that at first appear controversial, the resolution came when extreme views were attributed to their respective domains of validity. As is commonly the case, the matter is not black or white but rather degrees of gray.

---

1. In this book, the words *resource* and *consumer* are used interchangeably with *prey* and *predator*, respectively.

An excellent overview of predator-prey theory was published by Holt (2011), who presents the classical approach stemming from the Lotka-Volterra model as well as alternative views such as ours. The predominant standard theories are illustrated comprehensively by the books of Murdoch et al. (2003) and Turchin (2003). In this volume, we defend contrarian positions, with a goal that is much more focused on the axiomatic foundations of predator-prey theory than on a general review of the field. We attempt to show that the standard null model of elementary trophic interaction (the prey-dependent model) can be replaced by an equally simple but more realistic construction (the ratio-dependent model). While actual species interactions may lie somewhere "in the middle", the successive chapters of this book show that, in most cases, they are likely to be close to the proposed form and significantly far from the standard form. This observation, like any change of a basic assumption in a theory, has far-reaching consequences in predicting how food chains and food webs behave.

In this chapter, we start by pointing out the first clear appearance of the contradiction in microbiology in the 1940s and 1950s. We then explain briefly the analogous distinction that must be made in ecology between prey-dependent and ratio-dependent models. This is followed by the intermediate, predator-dependent models, which typically with one additional parameter span the space between the two extreme views. Our gradual interference hypothesis explains the whole picture by referring to consumer density. When density is low, consumers do not interfere and prey dependence works (as in the standard theory). When consumer density is sufficiently high, interference causes ratio dependence to emerge. In the intermediate densities, predator-dependent models describe partial interference. The next three chapters present the evidence supporting all our claims. Chapter 1 therefore serves as an exposition of our views without the proofs.

1.1 MONOD VERSUS CONTOIS: RESOURCE-DEPENDENT AND RATIO-DEPENDENT BACTERIA

For a long time, microbiology was a scientific domain far away from ecology. The two scientific communities of microbiologists and ecologists had little contact. However, during the past 10 to 20 years, there has been an increasing awareness that the dynamics of populations and communities of microorganisms follow by and large the same rules as those of macroorganisms (e.g., Bohannan and Lenski 1997; Kaunzinger and Morin 1998; Mounier et al. 2008). In our view, microbiologists have been ahead in many respects and ecologists have much to learn from them. The advance of microbiology can easily be explained by the economic demands of the food and drug industries. Studies of bacterial population growth made fast progress in the late nineteenth century and the first half of the twentieth century, both experimentally and theoretically, including the development of mathematical models.

Because bacteria come in very large numbers, their population size can be represented mathematically as a continuous variable. Exponential growth occurs in the situation of a well-mixed population with unlimited resources. In such a situation, the specific growth rate is a constant parameter, r_p, independent of both the bacterial density and the nutrient concentration. The growth of population size P is therefore represented by the differential equation

$$\frac{1}{P}\frac{dP}{dt} = r_p \tag{1.1}$$

whose solution is $P(t) = P_0 \exp(r_p t)$.

Of course, this fundamental law of population growth is only valid in ideal conditions. With a limited amount of resources, the growing population will reduce the resource concentration; the realized growth rate will decline; and, eventually, the population will stop growing. Over time, the population describes a sigmoid growth curve. At the beginning of the twentieth century, microbiologists first attempted to apply Verhulst's so-called logistic equation (see the historical account of Hutchinson [1978, 21–25]):

$$\frac{1}{P}\frac{dP}{dt} = r_p\left(1 - \frac{P}{K_P}\right) \tag{1.2}$$

Here, the carrying capacity K_P is the maximum size of the population and the parameter r_p is now the maximum specific growth rate, the rate at which the population would grow in the absence of density effects. Both parameters are asymptotic concepts that cannot be measured directly. The logistic equation is a convenient shortcut to represent the resource-limited growth of a population in situations when one is unable to represent explicitly the dynamics of its resource. It is widely used in ecology for this purpose.

As part of his doctoral work, French microbiologist Jacques Monod performed detailed experiments of bacterial growth. He showed that an excellent description of population dynamics could be given as the explicit combined result of (1) the decline of the nutrient concentration S and (2) the effect of this rarefied resource on the bacterial specific growth rate. Regarding the latter effect, he found that the specific growth rate was reduced when the nutrient concentration S was low. He proposed the following hyperbolic expression to describe this relationship (Monod 1942):

$$\frac{1}{P}\frac{dP}{dt} = \frac{r_p S}{A+S} \qquad (1.3)$$

The parameter r_p is the maximum growth rate, reached asymptotically in a situation of superabundant nutrients ($S \to \infty$). The parameter A is the so-called half-saturation constant: if the nutrient concentration is $S = A$, the realized specific growth rate on the right of eq. (1.3) becomes $r_p/2$. It was soon realized that the Monod function (1.3) was identical to the Michaelis-Menten law of enzyme kinetics. In ecology, this relationship was invented again, with a different parameterization, by Holling (1959b) to describe prey consumption by an individual predator. We return to this later.

Expression (1.3) establishes the link between the nutrient concentration S and the bacterial specific growth rate. It does not imply directly that the population should follow a sigmoid growth curve. If the nutrient is constantly maintained at some concentration S (e.g., in a chemostat), eq. (1.3) still says that the bacterial population grows exponentially, albeit at a reduced rate. In order to describe the full phenomenon of sigmoid growth, it is necessary to couple relationship (1.3) with the description of the decline of the nutrient concentration mentioned above.

Monod proved that there exists a proportional relationship between the quantity of bacteria created and the quantity of nutrient consumed. That is, nutrient mass is converted into bacterial biomass with energetic efficiency e. Therefore, the same function can be used to describe the nutrient uptake and the bacterial growth. The resulting system for the coupled dynamics of the nutrient and the bacterial biomass is

$$\begin{aligned} \frac{dS}{dt} &= -\frac{1}{e} \cdot \frac{r_p SP}{A+S} \\ \frac{dP}{dt} &= \frac{r_p SP}{A+S} \end{aligned} \qquad (1.4)$$

Monod showed that the mathematical solution $P(t)$ of eqs. (1.4) is a sigmoid curve qualitatively comparable to the solution of the logistic equation (1.2). However, he also showed that his experimental results are better described by the coupled system (1.4) than by the logistic equation (Monod 1942, 123–127).

Monod's main interest was to establish the way in which the nutrient concentration S affected the bacterial specific growth rate. In his experiments, he used well-mixed batch cultures with very low bacterial densities, focusing on the effect of different levels of nutrient concentration. He did not even consider higher densities that could affect the bacteria's own specific growth rate.

Left: Jacques Monod (b. Paris 1910; d. Cannes 1976). The early work of this French microbiologist explored the growth of bacteria with a quantitative approach: he proposed the "Monod function" in his doctoral thesis. He later became widely recognized as one of the founders of molecular biology and was awarded a Nobel Prize in 1965 with François Jacob and André Lwoff for discoveries concerning genetic control of enzyme and virus synthesis. Spending almost all his professional life at the Institut Pasteur in Paris, he also made a number of contributions to philosophy of science. See Wikipedia for additional historical information. Right: David Ely Contois (b. 1928; d. Honolulu 1988). Building on the results of Monod, this American microbiologist showed that the growth rate of bacteria declined with bacterial density. He developed the ratio-dependent "Contois function" in his doctoral thesis, prepared at the Scripps Institution of Oceanography at La Jolla, California. He then became a professor at the University of Hawaii, where he made his career. Drawings by Ksenia Golubkov.

This question was later addressed by American microbiologist David Ely Contois. As part of his doctoral research, he worked with organisms comparable to those of Monod (1942) but using a chemostat, the continuous-culture device that had been invented in the meantime (independently by Monod [1950] and by Novick and Szilard [1950]). This made possible a major difference: Contois worked with much higher bacterial densities. He found that the so-called half-saturation constant A was actually not a constant but was approximately proportional to the bacterial density P: $A = BP$. Consequently, instead of eq. (1.3), he suggested the following model for bacterial growth (Contois 1959):

$$\frac{1}{P}\frac{dP}{dt} = \frac{r_P S}{BP + S} \qquad (1.5)$$

or, equivalently,

$$\frac{1}{P}\frac{dP}{dt} = \frac{r_P}{B} \frac{S/P}{B + S/P} \qquad (1.6)$$

In the Contois model (1.6), the variable that determines the bacterial specific growth rate is the ratio of nutrient-to-consumer densities S/P, while in the Monod model (1.3) the growth rate is determined by the absolute nutrient concentration S. Contois did not give definite biological explanations for the differences from Monod's model. He speculated that the growth process was inhibited by soluble metabolic by-products released by the bacteria in proportion with their density P. Another plausible interpretation is the existence of a local heterogeneity around the cells due to their rapid uptake of nutrients (Lobry and Harmand 2006). Whatever the precise mechanism, Contois's experiments demonstrate that, in this range of densities, the bacteria interfered with each other so strongly that the specific growth rate was determined by the per capita availability of nutrients, S/P, not by the absolute nutrient concentration, S, as in Monod's model.

Today, the Contois model is used in various technological applications where it has been found to describe in a superior way both aerobic and anaerobic biodegradations, ranging from solid municipal organic waste to dairy manure and various industrial wastewaters. In fact, the Contois model is used as a default growth-rate model in simulations of the cleaning of wastewaters by microorganisms (a brief review is available in the introduction of Nelson and Holder [2009]). In general, the Monod model is used in ideal laboratory conditions of slow growth and low densities of pure strains, usually for research experiments of evolutionary biology, in which the population must be maintained in the exponential regime. In contrast, the Contois model applies widely to real-life "dirty" situations characterized by high densities, mixed bacterial communities, heterogeneous substrates, and so on. On the anecdotal side, Contois's article, before being published in the *Journal of General Microbiology*, had first been rejected by the *Journal of Bacteriology* with the editor's comment that, if it were correct, then Monod's formulation must be wrong, which could not be considered (Leslie Ralph Berger, personal communication).

Model developments in microbiology preceded homologous developments in ecology by 15 to 30 years (see Jost [2000] for a review of several parallels). As already mentioned, the Monod model was later reinvented by Holling (1959b). As for the Contois model, it was only recently realized that it was equivalent to the Arditi-Ginzburg ratio-dependent functional response (see section 1.3), with a different parameterization (e.g., Harmand and Godon 2007). An interesting point, which should be instructive in the ecological context, is that the Contois model of bacterial growth (despite being rejected by a first journal) was rather easily accepted by microbiologists, as applicable to different conditions from those of the already-established Monod model. In contrast, our homologous ratio-dependent model was received with skepticism by a number of theoretical ecologists, who generally stay loyal to the standard predator-prey model (i.e., the Lotka-Volterra model with its Rosenzweig-MacArthur variant; see section 1.2). We think that this difference in attitudes is explained by the fact that theory and application are much more divorced in ecology than they are in microbiology. Microbiologists accept pragmatically the theoretical developments that result from experiments. However, because it is not easy to set up real-scale well-controlled experiments involving large organisms, theoretical ecology has developed as a more or less closed discipline, with theoretical predictions often taking the status of facts. An example is the paradox of enrichment, commonly presented as a natural phenomenon. It is in fact only a mathematical consequence of prey dependence, with hardly any empirical support (see section 3.5). We return to the controversy around ratio dependence in chapter 5. However, the bulk of this book is devoted to a number of empirical and theoretical considerations which illustrate that, in most cases, the standard model makes wrong predictions while those of the ratio-dependent model are much more reasonable. We also attempt to delineate the domain in which the standard model remains valid and we propose a new versatile model that, with one additional parameter, can cover both types of situations.

1.2 THE STANDARD PREDATOR-PREY MODEL OF ECOLOGY

It is not this book's purpose to provide a general introduction to population dynamics and predation theory. Such an introduction can be found in every general textbook on ecology. We only review the minimal features of predator-prey models that are necessary to set the notations and to understand our proposed models. Nor do we attempt to review the very abundant research literature on predation theory. We refer to other authors' research articles only when needed for discussing the consequences of our developments.

The general framework was set independently by Lotka and by Volterra some 85 years ago with the system of differential equations

$$\frac{dN}{dt} = rN - aNP$$
$$\frac{dP}{dt} = eaNP - qP$$
(1.7)

where N is the prey abundance, P the predator abundance, r the specific prey growth rate, q the predator death and maintenance rate, and e the ecological efficiency with which prey consumed are converted into predator biomass. The central assumption of the model is the interaction term aNP, which appears in both equations, first to quantify the prey killed by the predators, and second to quantify the creation of predators. This term is chosen on the strength of an analogy with the law of mass action in chemistry, the idea being that, like an agitated solution of reacting molecules, predators and prey encounter each other randomly, the number of encounters being proportional to the abundance of both species. This mechanism was stated explicitly by Volterra (1931, 14). It is well known that the solutions of eqs. (1.7) are periodic functions that depend on initial conditions. Graphically, the states (N,P) describe a family of closed orbits in the phase plane $N \times P$.

The term aN can be interpreted as the number of prey killed by a single predator in a unit of time. Following Holling (1959b), it can be replaced by a more general function $g(N)$ called the *functional response*. Holling attributed deviations of the simple proportional relationship to additional complexities in patterns of behavior. For example, prey handling time leads to the so-called disc equation (Holling 1959b). As mentioned earlier, Holling's model is, in fact, identical to the Monod and the Michaelis-Menten models. With a predator-prey interpretation, it can be written as

$$g(N) = \frac{aN}{1+ahN} \qquad (1.8)$$

The *handling time*, h, is the time spent by the predator handling each prey encountered, and during which it stops searching. The parameter a, called *searching efficiency* (also called *attack rate* by some authors), can be interpreted as the proportion of prey killed per predator per unit of searching time.

Many other functional response models have been proposed but Holling's model is by far the most commonly used. Its main advantage over Lotka-Volterra's proportional model is that the functional response saturates hyperbolically to the value h^{-1}: as prey abundance N increases, a predator individual will not increase its kill rate indefinitely. Actually, rather than interpreting the parameter h in the strict behavioral sense of the time needed for handling a single prey item, it is best interpreted as the reciprocal of the asymptotic value of the functional response.

Another problem with the Lotka-Volterra model is the assumption of exponential prey growth in the absence of predators. The term rN can be replaced by some density-dependent function $F(N)$ like the logistic model

$$F(N) = r\left(1 - \frac{N}{K}\right)N \qquad (1.9)$$

The standard general predator-prey model can be written as

$$\frac{dN}{dt} = F(N) - g(N)P$$
$$\frac{dP}{dt} = eg(N)P - qP \qquad (1.10)$$

We have named this model *prey dependent* (Arditi and Ginzburg 1989) because the functional response g is a function that depends on prey density only:

$$g = g(N) \qquad (1.11)$$

This model, particularly with expressions (1.8) and (1.9) for $g(N)$ and for $F(N)$, was studied by Rosenzweig and MacArthur (1963) and Rosenzweig (1969, 1971, 1977), and its main features are presented in most textbooks. It can be analyzed graphically by the zero isocline method (figure 1.1). The predator isocline, obtained by setting the second of eqs. (1.10) to zero, is always vertical. The prey isocline, obtained by setting the first of eqs. (1.10) to zero, is an inverted parabola that crosses the N-axis at carrying capacity K. It can be "humped" or not in the positive quadrant. There is no hump if the searching efficiency a is relatively low. (It is in the negative quadrant.)

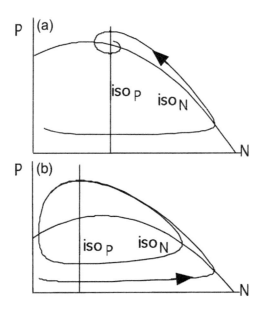

Figure 1.1. Isocline patterns in the Rosenzweig-MacArthur model. As in all prey-dependent models, the predator isocline is always vertical. (a) If the intersection lies on the right of the hump of the prey isocline, populations describe an inward spiral, converging to a stable equilibrium. (b) If predators are more efficient, the intersection can lie on the left side of the hump, giving rise to a limit cycle that attracts all trajectories, both from inside and from outside. From Jost (1998), with his kind permission.

The humped case must be considered generic since it includes all dynamic behaviors of the nonhumped case. The equilibrium is stable if the isoclines intersect on the descending part of the prey isocline. The equilibrium is unstable and gives rise to a limit cycle if the isoclines intersect on the ascending part of the prey isocline. This cycle is huge, spanning almost the whole interval $[0,K]$ for the prey.

The fact that the predator isocline is vertical has the important consequence that its abscissa alone determines the equilibrium value of the prey density. In the case of Holling's model (1.8), it is

$$N^* = \frac{q}{a(e-hq)} \tag{1.12}$$

Thus, prey equilibrium is set by four parameters that are present in the second equation of (1.10). It is entirely determined by the upper trophic level (perfect top-down control). In other words, prey equilibrium abundance does not depend in any way on its own demographic characteristics r and K, which are present only in the first equation of (1.10). This is a very unrealistic feature that brings a strong argument against prey-dependent models. We discuss this question in detail in this book, particularly in chapter 3.

For increasingly efficient predators (higher values of a or e, lower values of h or q), the predator isocline moves to the left. For enriched prey (higher value of K), the descending branch of the prey isocline moves to the right. Either of these changes gives rise to the so-called paradox of enrichment. An initially stable system (with the isocline intersection at the right of the hump, as in figure 1.1a) becomes destabilized if the prey's resources are enriched, or if the predators become increasingly efficient, because the isocline intersection can move to the left of the hump (as in figure 1.1b). We discuss this paradox in chapter 5, arguing that there is little evidence that it exists in nature.

1.3 THE ARDITI-GINZBURG RATIO-DEPENDENT MODEL

The rationale on which the standard model was developed by Volterra, Holling, Rosenzweig, and others never appeared convincing to us. The analogy with the law of mass action of chemistry rests on the "instantist" view that predation, reproduction, and mortality are truly continuous processes, all operating on the same time scale in a perfectly mixed environment (see chapter 5 for further explanation of what we call instantism). We hold that the functional response $g(N)$ in eqs. (1.10) should not be justified with a mechanism occurring on the behavioral time scale because these equations describe population dynamics, which operate on the slower time scale of reproduction and mortality. On this scale, the functional response may be completely different from the behavioral response. It must account in a simple macroscopic way for the intricacies of the predation process. Because of the many spatial and temporal heterogeneities that are present on the microscopic scale, various phenomena can induce predator density dependence in the feeding rate when calculated on the generation scale, even if predators do not interfere directly.

Taking the familiar example of lynxes and hares, eq. (1.8) means that the number of hares consumed per lynx in a unit of time would remain unchanged if the lynx population were doubled with the number of hares unchanged. On a daily time scale, the individual feeding rate can probably be modeled as the result of random encounters of hares by lynx individuals hunting independently from one another. However, when measured on the yearly time scale of population dynamics, it becomes natural to expect that the feeding rate must take account of predator abundance: over a year, fewer hares will be available for each lynx if the number of lynxes is doubled with the number of hares unchanged. Whatever the behavioral mechanisms are, the final outcome must reflect the fact that, with a given number of prey N, the available food is shared among the predators P and each individual's ration is reduced when more predators are present. This suggests that the yearly consumption rate should be a function of prey abundance per capita and not simply a function of the absolute prey abundance:

$$g = g\left(\frac{N}{P}\right) \qquad (1.13)$$

Of course, the heuristic considerations of the last two paragraphs are not proofs of the ratio-dependent response (1.13). For the moment, the latter should rather be considered as a working hypothesis, a null model. Chapters 2, and 3 provide ample empirical evidence that, in natural conditions, this is indeed the prevailing way in which the individual ration g depends on the population densities N and P. Chapter 4 presents a number of mechanisms showing how ratio dependence emerges, particularly from spatial heterogeneity and from time scale arguments.

Accepting the ratio-dependent model (1.13), we expect this function to be of the general shape of figure 1.2. With the ratio N/P as abscissa, the per capita consumption rate first increases at low ratios, but must be limited to some upper value when food is superabundant. In analogy with Holling's prey-dependent model (1.8), the ratio-dependent model that we suggest, which has become known as the Arditi-Ginzburg model, is the following:

$$g\left(\frac{N}{P}\right) = \frac{\alpha N/P}{1+\alpha h N/P} = \frac{\alpha N}{P+\alpha h N} \qquad (1.14)$$

The parameter h has the same units as in Holling's function (a time). As in that case, it is best interpreted geometrically as the reciprocal of the upper asymptote. As for the parameter α, it is not the same as the searching efficiency a of the prey-dependent model; it has different units. In both models, these parameters are, graphically, the slope of $g(\cdot)$ at the origin but, since the variable on the abscissa is not the same, the units are not the same. The parameter α can be interpreted as the rate at which the prey population is made available to the predator population: whatever the abundance of predators, the prey death rate due to predation cannot exceed α. While the parameter a of the prey-dependent model has some problems with units (it has the units of a surface when prey abundance is measured as a density), the parameter α of the ratio-dependent model has unambiguous units, irrespective of the fact that prey are measured in numbers, in density, or in biomass.

We coined the term *ratio dependence* (as well as prey dependence and predator dependence) in our first joint article on this subject (Arditi and Ginzburg 1989). However, this was not the first time that the ratio-dependent hypothesis was put forward. We had published the same idea, independently, several years earlier (Ginzburg et al. 1971, 1974; Arditi et al. 1977, 1978; Ginzburg 1983, 1986). In our 1989 article, we also mentioned an article by Getz (1984) that made very similar suggestions, and which was itself a theoretical elaboration of the applied simulation models of Gutierrez based on a supply-demand idea (Gutierrez and Baumgaertner 1984; see also Gutierrez 1996 and citations therein). Why didn't these earlier articles raise as much interest as ours? We believe that the main reason is that in the 1989 article and in subsequent work we explored the many community-level consequences of a change in the elemental trophic model, and we showed how predictions of the new theory could be confronted with readily

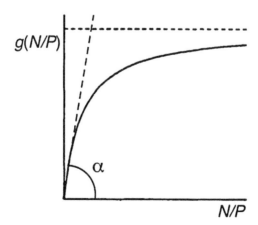

Figure 1.2. The ratio-dependent functional response is expected to be concave and monotonic, characterized by an initial slope α and an upper asymptote.

available observations, either dynamic or equilibrial. Such confrontations between models and data are the subjects of chapters 2, and 3.

When the Arditi-Ginzburg functional response (1.14) is introduced into the framework (1.10) together with logistic prey production, the following ratio-dependent dynamic model is obtained:

$$\frac{dN}{dt} = r\left(1 - \frac{N}{K}\right)N - \frac{\alpha N}{P + \alpha h N}P$$
$$\frac{dP}{dt} = e\frac{\alpha N}{P + \alpha h N}P - qP \quad (1.15)$$

The mathematical analysis of this model was first performed by Jost et al. (1999) and, more completely, by Berezovskaya et al. (2001). The solution trajectories present a rich range of patterns. The system can produce stable coexistence, limit cycles, deterministic predator extinction, and deterministic extinction of both species. Most of the biological interpretations can be done graphically with consideration of the isocline patterns only (figure 1.3).

The major geometric difference with the standard prey-dependent model is the fact that the predator isocline is now a slanted straight line passing through the origin:

$$\frac{N}{P} = \frac{q}{\alpha(e - hq)} \quad (1.16)$$

Passing through the origin has caused a variety of misunderstandings, and we address this question in section 1.6.

The prey isocline has different shapes depending on the comparison between α and r. If $\alpha < r$, the predation efficiency is limited and the prey equilibrium cannot drop lower than the value $N_L = K(1 - \alpha/r)$ shown on

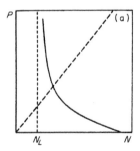
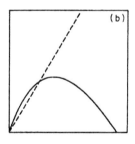
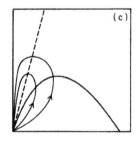

Figure 1.3. Isocline patterns of the ratio-dependent model. The predator isocline is always a slanted line through the origin. (a) If $\alpha < r$, the prey isocline has a vertical asymptote at N_L; the equilibrium point is always stable. (b) If $\alpha > r$, the prey isocline is always humped and the equilibrium point may be stable or unstable. (c) For very efficient predators, the only equilibrium point is the origin, which is a global attractor; typical trajectories are presented. From Arditi and Ginzburg (1989), with permission of Elsevier.

figure 1.3a. In this case, the equilibrium is always stable. If $\alpha > r$, the prey isocline is humped and passes through the origin (figure 1.3b). If the predator isocline is not steep, both populations can stably coexist. With a steeper isocline (more efficient predators), the intersection can lie on the ascending part of the prey isocline. With some additional conditions, this can give rise to limit cycles (see Jost et al. [1999] and Berezovskaya et al. [2001] for exact conditions). For even more efficient predators, the isoclines intersect only at the origin, which becomes a higher-order global attractor (figure 1.3c): the predators die out after exhausting the prey. This very reasonable outcome of predation, routinely observed in the laboratory (e.g., Gause 1934a, 1934b, 1935a), cannot be described by the standard prey-dependent model.

Note that the limit cycles that can appear are much smaller than those of the Rosenzweig-MacArthur model. They are also restricted to a very narrow range in the parameter space. Thus, predator-prey oscillations, while possible, are not a typical outcome of the ratio-dependent model. Except for the handful of spectacular cases that are reported in all textbooks (e.g., the Canadian lynx time series), oscillations are not a typical outcome of predator-prey interactions in nature. This is further discussed in section 5.4.

Comparing eqs. (1.16) and (1.12), one can notice that the same expression that determined the prey equilibrium biomass in the prey-dependent model now determines the equilibrium ratio. The parameters of the upper level determine the proportions of the population abundances at equilibrium but the absolute values also depend on the prey isocline, which depends on the parameters of the lower level. In other words, both population equilibria depend on the biological characteristics of both species. This is a much more realistic situation than in the standard model.

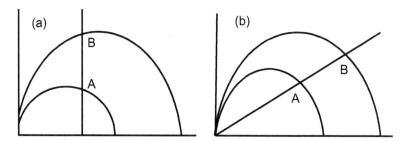

Figure 1.4. Response to prey enrichment from A to B. (a) In the standard prey-dependent model, enrichment increases the predator equilibrium abundance but not the prey equilibrium; the whole system can shift from stable to unstable if the isocline intersection passes from the right to the left of the hump, as in this example. (b) In the ratio-dependent model, both population equilibrium abundances respond to enrichment; if the system is initially stable, it remains stable after enrichment.

The ratio-dependent model is not destabilized by prey resource enrichment. Figure 1.4b shows that, in response to an increase of the prey carrying capacity, both the prey and the predator equilibria increase in a proportional manner. If the equilibrium point is initially stable, it remains stable after enrichment. Again, this seems to be in agreement with reality (see chapter 3 and section 5.2).

1.4 DONOR CONTROL AND RATIO DEPENDENCE

Because of the assumed concavity of $g(\cdot)$ (see figure 1.2), the ratio-dependent functional response obeys the following inequality:

$$g\left(\frac{N}{P}\right) \leq \alpha \frac{N}{P} \tag{1.17}$$

or

$$g\left(\frac{N}{P}\right) P < \alpha N \tag{1.18}$$

The expression on the left of eq. (1.18) is the total amount of prey killed by all predators. It cannot exceed αN and it tends to this value for small N or large P. In other words, the amount of prey made available to the whole predator population is only determined by the prey density.

If saturation is neglected in the ratio-dependent model (1.14), that is, if $h = 0$, the dynamic system (1.15) becomes

$$\begin{aligned}\frac{dN}{dt} &= rN\left(1 - \frac{N}{K}\right) - \alpha N \\ \frac{dP}{dt} &= e\alpha N - qP\end{aligned} \tag{1.19}$$

In these equations, the total number of prey killed becomes exactly equal to αN. Thus, the number of prey consumed as well as the number of predators produced are simply proportional to prey abundance; they do not depend on predator abundance. Following Pimm (1982), we name this situation *donor control* because the prey (the donor) is the sole determinant of predator production. At the same time, the predator abundance P is absent in the prey equation. The predators impose a mortality rate on the prey, α, but this mortality is constant. If there are no predators, the predation term αN should be replaced by zero. So, technically, the predation term in the prey equation must be written $-g(N/P) \cdot P$ with:

$$g\left(\frac{N}{P}\right)P = \begin{cases} \alpha \dfrac{N}{P} P = \alpha N & \text{if } P \neq 0 \\ 0 & \text{if } P = 0 \end{cases} \quad (1.20)$$

Actually, the second case, $g(N/P)P = 0$ if $P = 0$, should be added to all ratio-dependent functional response models, including Arditi-Ginzburg (eq. 1.14) and others. For simplicity, we generally omit this addition, but it must be kept in mind.

Note that our model (1.19) is quite different from Pimm's original donor-control model (Pimm 1982, 16). In Pimm's model, the prey has quadratic mortality, independent of the presence of predators, and the predator equation is a ratio-dependent expression built after Leslie's model (1948), which suffers logical problems that we explain below (section 1.7). The only possible interpretation of Pimm's model is the very special case of predators that consume dead or dying individuals (such as hyenas, the example originally given by Pimm) and, thus, have no effect on the prey population. In our donor-control model (1.19), valid for a much more general situation, the presence of predators clearly has an effect on the prey dynamics, but this effect simply does not depend on predator density.

The donor control functional response model (1.20) is nothing other than the Arditi-Ginzburg ratio-dependent model for the case of insatiable predators. This intimacy between the two models is rarely recognized. Indeed, some authors who have opposed ratio dependence do not object to the donor control model, failing to realize that the latter is a special case of the former (e.g., Abrams and Walters 1996).

We can extrapolate the properties of our strict donor control definition given above to more general ratio-dependent functional responses of the type seen in figure 1.2. If prey and predator densities are such that predators are not satiated (i.e., the functional response remains at the left of the curve), then variations in predator numbers hardly alter the total mortality imposed upon the prey population. This implies a stabilizing influence: for example, a 50% decrease in predator numbers will not lead to a surge of prey numbers because the overall mortality will hardly change. This is completely different from prey-dependent systems. In chapter 3, we examine the consequence of this stabilizing influence in complex food webs based on donor control.

Conversely, if prey and predator densities are such that predators are satiated (i.e., the functional response remains at the right of the curve), both ratio-dependent and prey-dependent models predict that total prey mortality is independent of prey abundance and proportional to predator abundance. This is well known to be a source of instability because, say, a 50% increase in prey numbers will not be hindered by increased mortality.

1.5 PREDATOR-DEPENDENT MODELS

We have named *predator-dependent* the functional responses of the form $g = g(N,P)$, where the predator density P acts (in addition to N) as an independent variable to determine the per capita kill rate (Arditi and Ginzburg 1989). The specific ratio-dependent form (1.13) is the special case of perfect sharing, while the prey-dependent form (1.11) is the special case of no sharing. It is certainly a safe position to accept the existence of predator dependence as a general case, with the common argument that "nature is complex." However, adding more details to the description of nature is not necessarily desirable and must be justified. Mathematically, predator-dependent functional response models have one more parameter than the prey-dependent or the ratio-dependent models. Some examples of such models are given in table 1.1. The main interest that we see in these intermediate models is that the additional parameter can provide a way to quantify the position of a specific predator-prey pair of species along a spectrum with prey dependence at one end and ratio dependence at the other end:

$$g(N) \leftarrow g(N,P) \rightarrow g\left(\frac{N}{P}\right) \qquad (1.21)$$

In the Hassell-Varley and Arditi-Akçakaya models (see table 1.1), the mutual interference parameter m plays the role of a cursor along this spectrum, from $m = 0$ for prey dependence to $m = 1$ for ratio dependence. Note that this theory does not exclude that strong interference goes "beyond ratio dependence," with $m > 1$.[2] This is also called *overcompensation*. A comparative analysis of the dynamic properties of these models was made by Arditi et al. (2004). They showed that in the normal, undercompensation range, interference has a favorable effect on stability and resilience. For overcompensation values, it is the contrary.

In this book, rather than being interested in the interference parameters per se, we use predator-dependent models to determine, either parametrically or nonparametrically, which of the ends of the spectrum (1.21) better describes predator-prey systems in general. It is of great interest to model

---

2. We use the word *interference* as a synonym for *predator dependence*. It generally designates not a behavioral mechanism but the fact that increasing predator density depresses the average individual predator food intake.

Table 1.1. THE FUNCTIONAL RESPONSE MODELS USED IN THIS BOOK

Prey-Dependent Models $g(N)$	Predator-Dependent Models $g(N,P)$	Ratio-Dependent Models $g(N/P)$
LV: aN (HV with $m = 0$)	HV: $\dfrac{\alpha N}{P^m}$	AG-DC: $\dfrac{\alpha N}{P}$ (HV with $m = 1$)
General prey dependent: $g(N)$ (GHV with $m = 0$)	GHV: $g\left(\dfrac{N}{P^m}\right)$	AG, general ratio-dependent: $g\left(\dfrac{N}{P}\right)$ (GHV with $m = 1$)
Ho: $\dfrac{aN}{1+ahN}$ (AA with $m = 0$)	AA: $\dfrac{\alpha N P^{-m}}{1+\alpha h N P^{-m}}$	AG type II: $\dfrac{\alpha N/P}{1+\alpha h N/P}$ (AA with $m = 1$)
Tends to Ho for $P \ll P_c$	Gradual interference: $\dfrac{aN}{P/P_c + \exp(-P/P_c) + ahN}$ (Tyutyunov et al. 2008)	Tends to AG for $P \gg P_c$
Becomes identical to Ho for $c = 0$	DAB: $\dfrac{aN}{1+ahN+cP}$	Tends to AG for $ahN + cP \gg 1$
Monod, equivalent to Ho: $\dfrac{g_{max} N}{A+N}$ (with $B = 0$)	DAB with different parameterization: $\dfrac{g_{max} N}{A+N+BP}$ (Ruxton and Gurney 1992)	Contois, equivalent to AG: $\dfrac{g_{max} N}{N+BP}$ (with $A = 0$)
$g_{max}\left(1-e^{-cN}\right)$ (Ivlev 1961; Watt with $m = 0$)	$g_{max}\left(1-e^{-cN/P^m}\right)$ (Watt 1959)	$g_{max}\left(1-e^{-cN/P}\right)$ (Gutierrez and Baumgaertner 1984; Watt with $m = 1$)

The first four rows contain those models that are most central to the book. Models on the same row contain the predator-dependent expression in the center, and the corresponding prey-dependent and ratio-dependent limits.

AA, Arditi-Akçakaya 1990; AG, Arditi-Ginzburg 1989; AG-DC, Arditi-Ginzburg with donor control*; DAB, DeAngelis et al. 1975 or Beddington 1975 (with some approximation); Ho, Holling type II; HV, Hassell-Varley 1969; GHV, generalized Hassell-Varley (Arditi, Ginzburg, and Akçakaya 1991); LV, Lotka-Volterra.

*Substantially different from Pimm's commonly used definition of donor control (see section 1.3 for details and section 3.6 for consequences in trophic web structure).

predation with a simple expression accepted as the null model. It is needed in theoretical studies of general ecosystem properties, like food web dynamics. A simple expression is also needed in applied simulation models of ecosystems involving numerous consumer-resource links or in conservation studies of specific populations. If necessary, complications can be added to the starting model but, obviously, the starting model should be the one that presents the best approximation to the natural system.

1.6 WHAT HAPPENS AT LOW DENSITY? THE GRADUAL INTERFERENCE HYPOTHESIS

The ratio-dependent model has repeatedly been criticized for the fact that the predator isocline passes through the origin (e.g., Abrams 1994; Gleeson 1994). Indeed, this shape means that, for any arbitrarily small prey density (externally maintained), a predator population can theoretically exist in equilibrium with it. This predator population would be small, in commensurate proportion with the small prey population, but nonzero. Some authors consider that, because of the maintenance costs of a single individual, no predator population should be allowed to exist below a distinct threshold of prey density. These authors therefore advocate for a predator isocline that intersects the prey axis at some positive value.

It is a legitimate question to ask whether the implied prey sharing of ratio dependence still occurs at low predator density. Taking again our pet example of lynxes and hares, what happens if the lynx density is so low that the home ranges of different individuals never overlap, even over a whole year? Clearly, in such a situation, each individual lynx must feel virtually alone in the environment. In other words, the functional response should be prey dependent. This is equivalent to saying that the predator isocline should be vertical at low density. Still, at high densities, the functional response should become ratio dependent for the reasons given in section 1.3. Further considerations on the effect of home ranges have been proposed by Ginzburg and Jensen (2008) and are explained in chapter 4.

Thus, it is quite conceivable that the predator isocline could have one of the shapes shown in figure 1.5. It could be vertical at low densities, becoming gradually more slanted and aligned with a line passing through the origin. The case of bacterial growth that we discussed in section 1.1 also favors this view. As we pointed out, Contois's ratio-dependent growth function was observed on species similar to those on which Monod had established his resource-dependent function. The only important difference was bacterial density: low densities presented resource-dependent growth; high densities presented ratio-dependent growth.

While the purely ratio-dependent isocline is characterized by a single parameter (its slope), the present gradual isocline has two essential parameters: the position of the vertical part and the slope of the slanted part.

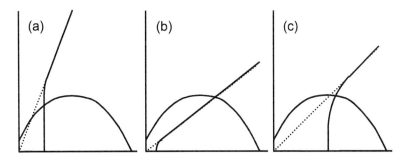

Figure 1.5. Possible isocline patterns in a gradual interference model. The predator isocline is vertical at low predator density and slanted through the origin at higher density. Depending on the parameter values, (a) prey dependence can predominate; (b) ratio dependence can predominate; or (c) an intermediate predator dependence can predominate in the vicinity of the equilibrium.

The development of such models with *gradual interference*, prey dependent at low density and ratio dependent at high density, is a recent line of research (Ginzburg and Jensen 2008; Tyutyunov et al. 2008; Tran 2009). In chapter 4, it is shown how gradual functional responses can emerge. One possible expression was proposed by Tyutyunov et al. (2008):

$$g(N,P) = \frac{aN}{P/P_c + \exp(-P/P_c) + ahN} \quad (1.22)$$

At low predator density $(P \ll P_c)$, the term $P/P_c + \exp(-P/P_c)$ in the denominator is close to 1 and the expression (1.22) tends toward the prey-dependent Holling model. At high predator density $(P \gg P_c)$, the term $\exp(-P/P_c)$ becomes close to 0 and the expression (1.22) tends toward the ratio-dependent Arditi-Ginzburg model, with the reparameterization $\alpha = aP_c$.

A functional response $g(N,P)$ like eq. (1.22) is predator dependent in the sense defined in section 1.5: the prey and predator densities act as two independent variables to determine the value of g. However, it is very different from all examples given in table 1.1. In those examples, the intensity of interference was quantified by some parameter (e.g., the parameter m). Biologically, a given predator-prey system would be characterized by a given value of this parameter, with interference taking place with the same (intermediate) intensity at all population densities. With a model like eq. (1.22), on the contrary, prey dependence and ratio dependence both occur in the same biological system, at the opposite extremes of predator densities. The parameter P_c quantifies the level of predator density at which ratio dependence gradually replaces prey dependence. The biological question becomes the estimation of the value of this parameter. If P_c is large in comparison with commonly occurring values of the predator density P, prey dependence would predominate (figure 1.5a); if it is small, ratio dependence would predominate (figure 1.5b).

Another way of posing the biological question is to ask what is the slope of the predator isocline in the region of prevailing predator densities (e.g., around the equilibrium). Since the isocline is curvilinear, figure 1.5c shows that the slope around the equilibrium can have an intermediate value between being vertical (prey dependence) and being aligned with the origin (ratio dependence). Thus, the predator-dependent intermediate models of section 1.5 can be interpreted as approximations of the gradual interference hypothesis in a given range of predator densities. In particular, the Arditi-Akçakaya model remains very useful because its mutual interference parameter m can be interpreted as the reciprocal of the slope of the predator isocline: it varies between $m = 0$ where the isocline is vertical and $m = 1$ where it is aligned with the origin.

In a situation with predominating ratio dependence (as in figure 1.5b), the simplification consisting of extending the slanted isocline all the way to the origin could be acceptable because population trajectories should not be interpreted literally in the vicinity of the origin. As in any model framed in differential equations, there is the assumption that population abundances can be represented by continuous variables. For this to be acceptable, populations must be made up of many individuals. In truly small populations, a different mathematical formalism must be used, with population abundances described as discrete variables, since the numbers of individuals come as integers. Demographic variations must also be modeled as stochastic processes, not as deterministic continuous functions, as is done in differential equations. Actually, the study of small populations belongs to a completely different body of theory, more commonly used in conservation ecology than in community ecology. Clearly, our book is set in the latter body of theory. We therefore assume that low population densities are still sufficiently high for the populations of prey and predators to be reasonably represented by the continuous variables N and P. With this assumption, which is also made by Lotka-Volterra and other prey-dependent models, ratio dependence can still be considered at low densities and there is nothing pathological with the predator isocline going through the origin (Akçakaya et al. 1995). Moreover, the ratio-dependent model predicts that initially low populations would either grow to a higher stable equilibrium (figure 1.3a or 1.3b), or grow transitorily before going to joint extinction (figure 1.3c), or possibly enter oscillations on a stable limit cycle (which can occur in the situation of figure 1.3b). Thus, even though the predator isocline is present at low densities, the model does not say that the species can coexist in a stable low equilibrium.

1.7 BIOMASS CONVERSION

In all the above resource-consumer dynamic models (1.4, 1.7, 1.10, 1.15, 1.19), we wrote the production term in the consumer equation as identical to the consumption term in the resource equation (differing only by a constant factor e). More generally, let us write the resource-consumer system as

$$\frac{dN}{dt} = \text{resource production} - g(\cdot)P$$
$$\frac{dP}{dt} = h(\cdot)P \tag{1.23}$$

and let us focus on the relationship between the functions g and h. As a companion to the functional response g, the function h has traditionally been named *numerical response*. It is unfortunate that, when elements of models have names, people may think of them as separate, unrelated

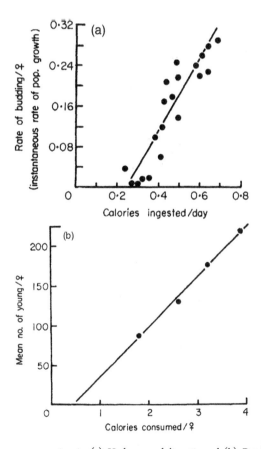

Figure 1.6. Biomass conversion in (a) *Hydra pseudoligactis* and (b) *Daphnia pulex*. From Beddington et al. (1976), with permission of Wiley-Blackwell.

entities. We give such examples shortly. However, it is logical to assume that the rate of consumer production h depends on the rate of consumption g, that is (Ginzburg 1998):

$$h = h(g(\cdot)) \quad (1.24)$$

This reflects the fact that consumers can only be produced from the resources they consume. Equation (1.24) can be called the rule of biomass conversion, which is not necessarily proportional—it can be nonlinear. While allowing for an arbitrary relationship, it is reasonable to think of $h(\cdot)$ as an increasing function, with the amount consumed on the abscissa. Two experimental illustrations are given in figure 1.6.

It follows from the rule of biomass conversion, generally nonlinear, that the predator production h does not respond directly to the population densities N and P; it responds directly only to consumption g. Since the functional response g is a function of the population densities, $g(N,P)$, the numerical response h can also be written as a function of the population densities, $h(N,P)$. However, it must be kept in mind that, biologically, the densities N and P determine the value of h only indirectly, through g. Therefore, the function $h(\cdot)$ must present the same type of dependence as the function $g(\cdot)$. If g is prey dependent, then h must be prey dependent. If g is ratio dependent, then h must be ratio dependent.

The model by Leslie (1948) violates the conversion principle. It uses a prey-dependent Lotka-Volterra functional response and a ratio-dependent numerical response, with the predators having a logistic-like growth rate, their own carrying capacity kN being set by the amount of prey:

$$\frac{dN}{dt} = r\left(1 - \frac{N}{K}\right)N - aNP$$
$$\frac{dP}{dt} = s\left(1 - \frac{P}{kN}\right)P \quad (1.25)$$

In retrospect, Leslie's model can be explained as an attempt to improve the Lotka-Volterra model, particularly with respect to the unrealistic properties of the vertical predator isocline. Leslie's predator isocline is a slanted line through the origin, exactly as in the ratio-dependent model (1.15). This means that, for any given (fixed) prey density, predators eventually equilibrate at a level that is proportionate with that of the prey. This realistic property of the slanted predator isocline can explain why Leslie's model has been used up to the present by some authors, most notably by May (1974, 84, 89–90, 189–193), by Berryman (1999; and references therein), and by Turchin (2003, 317–325; and references therein).

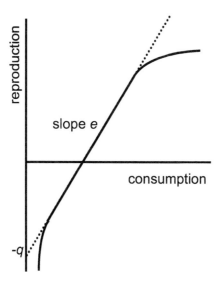

Figure 1.7. Hypothetical conversion between the per capita rates of prey consumption g and predator reproduction h (solid curve). The dotted line is a linear approximation at low to moderate consumption. After Ginzburg (1998), with permission of Wiley-Blackwell.

A problem with Leslie's model is that the improvement of the predator isocline is obtained in an inconsistent way because, in eqs. (1.25), there is no relationship between the rate at which predators reproduce and the rate at which prey are consumed. Since per capita consumption and per capita reproduction are functions of different arguments (consumption increases with N and reproduction increases with N/P), it is quite possible to choose any one of them (say, N) to be arbitrarily small and the other one (say, N/P) arbitrarily large. Consequently, a very high rate of reproduction would be generated by a very low consumption. This is certainly nonsensical, as predators would reproduce by "spontaneous generation." Moreover, as we explained in section 1.1, the fundamental rationale of the logistic model is to describe the dynamics of a resource-limited population in situations where the dynamics of the resource itself are not described. It is therefore inconsistent to couple, as in eqs. (1.25), the logistically growing predator with an equation for its resource.

Almost all predator-prey models that satisfy the conversion rule satisfy, in fact, a more restrictive assumption of linear conversion (as in equations 1.4, 1.7, 1.10, 1.15, 1.19):

$$h(\cdot) = eg(\cdot) - q \qquad (1.26)$$

where e is the conversion efficiency ($0 < e < 1$ when using biomass units for N and P) and with the usual interpretation of q as the maintenance costs and natural mortality of predators.

We can view (1.26) as the linear approximation of a more complex relationship between functional and numerical responses (figure 1.7). At very low food levels, it is conceivable that mortality could become infinitely high. At very high levels of consumption, additional consumption (possibly just killing prey without eating them) may not lead to proportionally more reproduction. The linear approximation is likely to be more accurate when predators are hungry. The examples of figure 1.6 lend strong empirical support to this approximation.

CHAPTER 2

Direct Measurements of the Functional Response

As explained in chapter 1, the fundamental problem of the Lotka-Volterra and the Rosenzweig-MacArthur dynamic models lies in the functional response and in the fact that this mathematical function is assumed not to depend on consumer density. Since this function measures the number of prey captured per consumer per unit time, it is a quantity that should be accessible to observation. This variable could be apprehended either on the fast behavioral time scale or on the slow demographic time scale. These two approaches need not necessarily reveal the same properties: as argued in chapter 1, it is conceivable that a given species could display a prey-dependent response on the fast scale and a predator-dependent response on the slow scale. The reason is that, on a very short scale, each predator individual may "feel" virtually alone in the environment and react only to the prey that it encounters. On the long scale, the predators are more likely to be affected by the presence of conspecifics, even without direct encounters. In the demographic context of this book, it is the long time scale that is relevant. A very nice illustration of this is provided by the long-term studies of Norman Owen-Smith on the kudu antelope in Africa. He established more than 20 years ago that annual survival and juvenile recruitment in this herbivore were functions of annual rainfall (governing plant production) relative to the population biomass, that is, a per capita relationship (Owen-Smith 1990). This ratio-dependent relationship appears when the demographic rates are viewed on the annual scale. On a daily time scale, food intake is simply determined by the current food availability; that is, it is prey dependent (Owen-Smith 2002, 222–224).

In this chapter, we review a number of experimental or observational studies that lead to direct quantitative estimates of the numbers of captures and of the way they depend on consumer densities. Some of these experiments were set on the fast behavioral scale, which is not necessarily appropriate. However, they can still be instructive because the effects of any behavioral interference cannot disappear on the slow scale. Therefore, if predator dependence is detected on the fast scale, then it can be inferred that it must be present on the slow scale; if predator dependence is not detected on the fast scale, it cannot be inferred that it is absent on the slow scale.

Since we are interested in the ways that the predator density affects the functional response, we do not review those studies in which the prey density was varied while not varying the consumer density. Actually, these make up the vast majority of experimental studies on the functional response, which only sought to parameterize the dependence on prey density. Following the work of Holling (1959a), a great number of studies were undertaken to identify the "type" (i.e., I, II, or III) or to estimate the attack rate a and the handling time h and the ways they vary or not in different conditions. Most of these studies were carried out with a single individual predator, often not considering the fact that the presence of several individuals might change the results.

Many short-scale behavioral studies of the influence of the consumer density were performed in the laboratory with insects or similarly small animals. They are reviewed in section 2.1. Most reveal a significant effect of consumer density on the functional response, the effect called *interference*. However, being focused on the question of the influence of consumer density, many studies did not compare the outcome at various prey densities. We show that this may lead to an underestimation of the strength of interference. Analyzing a number of experiments in which both population densities were varied, we show that most systems display nearly perfect compensatory interference, meaning that the individual searching efficiency is reduced in exact proportion to the predator density. This is equivalent to ratio dependence. Since some scientists have argued that the high level of interference observed in laboratory experiments may be artifactual, we then review evidence based on field observations, which are, of course, more rare and less precise. Section 2.2 is a detailed analysis of the only experiment ever set up to purposely test the ratio dependence hypothesis in the field. Using predatory wasps, it leads to a clear case of approximate ratio dependence.

The results of sections 2.1 and 2.2 pertain to behavioral-scale situations. With section 2.3, we show how a long-scale demographic study can bring up important information. This section discusses the analysis of a unique data set: the time series of wolves and moose on Isle Royale, an isolated predator-prey system for which the numbers of kills have been monitored in addition to the population numbers. Again, approximate ratio dependence appears to be the case. Finally, section 2.4 reports four previously published experimental studies performed with the explicit purpose of testing ratio dependence.

2.1 INSECT PREDATORS AND PARASITOIDS, SNAILS, FISH, AND OTHERS: LABORATORY MEASUREMENTS

2.1.1 Manipulating the Consumer Density Alone

In parallel to the development of predator-prey theory in the 1930s–1950s, an increasing number of laboratory experiments were set up to estimate the efficiency of predators and its dependence on various factors. Many such experiments have been performed with insects, often for applied purposes, such as the biological control of invading pests. Typically, prey and predators (or hosts and parasitoids) are left interacting in a closed arena for some duration T (e.g., 24 hours). The observation is the number of prey attacked N_a, for a given initial number of prey N and with a fixed number of predators P. The duration T is chosen short enough that natural birth and death events do not occur. Assuming the simple Lotka-Volterra "mass action" instantaneous interaction aNP, the total number of prey attacked in the course of T is given by the Nicholson-Bailey expression

$$N_a = N[1-\exp(-aPT)] \qquad (2.1)$$

Such a typical experiment gives a measurement of the searching efficiency a because eq. (2.1) leads to

$$a = \frac{1}{PT}\log\left(\frac{N}{N-N_a}\right) \qquad (2.2)$$

Under the assumptions of this model, a single interaction session (with known values of N, P, and T) is theoretically sufficient to estimate a.

Of particular interest to us are those experiments designed to investigate the effect of predator density on the searching efficiency. This effect can be due to direct behavioral encounters among predators, or to more complex spatial phenomena (e.g., aggregations) that make the searching efficiency decline with predator density. Whatever the mechanism, a convenient way to describe the magnitude of this effect was proposed by Hassell and Varley (1969) when they defined the mutual interference parameter m as the opposite of the slope of the log-log regression of observed values of a for increasing values of P:[1]

---

[1]. As explained in chapter 1, we use *interference* as a synonym for predator dependence. It designates the fact that predator density depresses the average individual food intake.

$$\log a = -m \log P + \text{const.} \qquad (2.3)$$

or equivalently

$$a = \alpha P^{-m} \qquad (2.4)$$

Because they used eq. (2.2) to estimate the searching efficiency and eq. (2.4) to describe the way it is dependent on predator density, Hassell and Varley effectively built the following predator-dependent functional response model, which we denote as HV in the rest of this book:

$$g(N, P) = \alpha N P^{-m} \qquad (2.5)$$

Hassell and Varley (1969), Hassell (1978, 2000, and references therein), and other authors published a number of values of m derived from experiments of the sort described above (e.g., figure 2.1). Most of these values are in the range 0.3–0.9. It is particularly noteworthy that values $m \approx 0$, which would characterize a Lotka-Volterra response $g(N)=aN$, were never observed. Negative values were observed very rarely; they would indicate cooperation rather than interference. Values higher than 1 have been observed, but are uncommon. They indicate overcompensating interference, that is, a situation in which the total consumption of prey by all predators $[g(\cdot)P]$ declines for increasing numbers of predators.

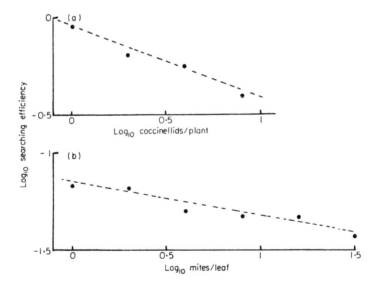

Figure 2.1. Two laboratory experiments illustrating the application of the HV model to estimate the mutual interference parameter. (a) The predatory ladybird *Coccinella septempuncta* feeding on aphids; $m = 0.38$. (b) The predatory mite *Phytoseiulus persimilis* feeding on mite prey; $m = 0.18$. From Hassell et al. (1976), with permission of Wiley-Blackwell. See also Hassell (1978, 82, 91) for biological details.

In the framework of model (2.5), ratio dependence is the special case $m = 1$: $g(N/P) = \alpha N/P$. We already encountered this model in chapter 1 (see particularly table 1.1). It describes donor control because the total prey consumption by all predators is αN, that is, it is solely determined by prey abundance, independently of the number of predators. Most studies applying the above procedure do not report values of m close to 1. However, in the next section we show that this procedure, ignoring the saturating nonlinearity of the functional response, tends to underestimate the magnitude of the interference parameter. When corrected, many values become close to 1.

2.1.2 Measuring Interference in the Presence of a Saturating Functional Response

In the previous section, we saw that, following the estimation procedure based on the HV model, most species display an interference coefficient m that is significantly different from zero. In the terminology introduced in chapter 1, the experimental evidence is that most predator-prey interactions are therefore not purely prey dependent but are predator dependent instead.

The HV model uses the Nicholson-Bailey expression (2.1) for the number of prey attacked N_a, implying that the interaction time T is entirely devoted to searching. The functional response is assumed to increase indefinitely (linearly) with increasing prey abundance. As already mentioned in section 2.1.1, the observation of N_a for a single value of N is enough to estimate the searching efficiency a because expression (2.2) can be extracted from (2.1).

However, if the predator presents a saturating functional response (so-called type II), for example because of a handling time h, expression (2.1) must be replaced by the following expression (Royama 1971; Rogers 1972):

$$N_a = N[1 - \exp(-aPT + ahN_a)] \qquad (2.6)$$

Note that this is an implicit expression for N_a because this variable appears on both sides of the equation and cannot be isolated. For a given value of P (and T), the observation of the number attacked N_a at a single value of N can no longer give an estimate of a since (2.6) also contains the unknown parameter h. Ideally, a complete functional response curve (N_a/P as a function of N) should be obtained for each value of P (see figure 2.2 for an example). For each of these curves, the nonlinear regression of expression (2.6) for N_a can provide estimates of a and h. Collecting

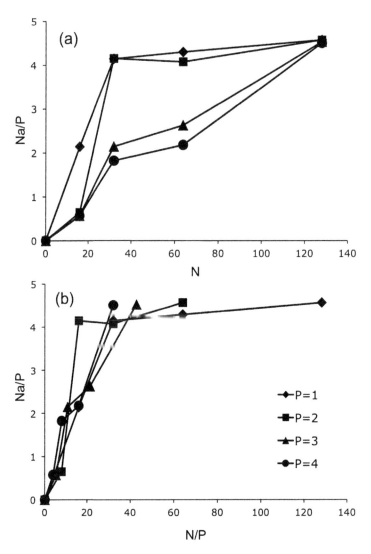

Figure 2.2. (a) Functional responses of marine snails preying on barnacles at four different snail densities. (b) The same data are plotted against the ratio N/P, illustrating approximate ratio dependence because the four curves become superimposed. Data from Katz (1985).

the estimates of a, the mutual interference coefficient m is finally obtained with a linear regression of $\log a$ against $\log P$.

If the searching efficiency is calculated with eq. (2.2) despite the fact that predators have a nonzero handling time, a bias is introduced. This bias is illustrated in figure 2.3. A saturating predator was simulated by using eq. (2.6) to calculate N_a for several values of P, with chosen values of

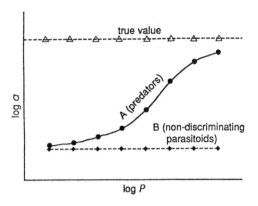

Figure 2.3. The use of the Nicholson-Bailey expression (2.2) to estimate the searching efficiency a gives biased values if the handling time is not zero. Here, the true value of a does not vary with P (no interference). Curve A shows the estimate of a when using eq. (2.2) instead of (2.6). Horizontal line B shows the estimate of a when using eq. (2.2) instead of (2.8). See text for additional explanations. From Arditi and Akçakaya (1990), with kind permission of Springer Science+Business Media.

N, a, h, and T. Using these values of N_a, expression (2.2) was then applied as an estimator of a. Even though the correct value of a was the same for all values of P (no interference), figure 2.3 (curve A) shows that the estimate of a using eq. (2.2) increases with P. Calculating m by fitting a straight line through these points would give a negative value (positive slope) instead of $m = 0$. The longer the handling time, the steeper this slope. If interference is present, the correct slope should be negative. Since the bias adds a positive contribution, it can be inferred that values of m resting on the use of eq. (2.2) are underestimated.

In the case of parasitoids, an appropriate attack equation was given by Arditi (1983):

$$N_a = N\left[1-\exp\left(\frac{-aPT+a(h-h_p)N_a}{1+ahN}\right)\right] \quad (2.7)$$

where h and h_p are handling times on encounters with healthy and parasitized hosts, respectively. In the special case where $h_p = 0$, this equation reduces to the predator equation (2.6): parasitoids that instantly recognize and abandon parasitized hosts behave like predators. In the case where $h_p = h$, parasitoids do not discriminate between healthy and parasitized hosts and eq. (2.7) reduces to the following expression (Rogers 1972):

$$N_a = N\left[1-\exp\left(\frac{-aPT}{1+ahN}\right)\right] \quad (2.8)$$

Figure 2.3 (curve B) shows the values of a given by the estimator (2.2) when N_a actually follows (2.8): the absolute magnitude of a is estimated incorrectly but the bias does not vary with P. Therefore, slope m is not biased as in the case of predators. In the case of imperfectly discriminating parasitoids, the bias has intermediate severity.

We can conclude that most published values of m, like those mentioned in section 2.1.1, are likely to be underestimated. The only ones that can be trusted are those pertaining to predators with negligible handling time or to parasitoids that are strictly nondiscriminating. Using real data, we will now compare the values of the interference constant m estimated with the Nicholson-Bailey equation (2.1) to values obtained with a corrected method based on eq. (2.6), a method in which parameter h is estimated along with searching efficiency a.

2.1.3 The Arditi-Akçakaya Predator-Dependent Model

Making the straightforward combination of the Hassell-Varley hypothesis (2.4) and the standard Holling model (1.8), Arditi and Akçakaya (1990) have proposed the following general model (AA) for describing a functional response that is dependent on both prey and predator densities (i.e., predator dependent):

$$g(N,P) = \frac{\alpha N P^{-m}}{1+\alpha h N P^{-m}} \qquad (2.9)$$

This model is largely used throughout the rest of this book. Its main interest is that it is general enough to cover a whole spectrum of responses with various intensities of predator dependence. Two special cases are particularly important. For $m = 0$, this model reduces to the usual Holling prey-dependent model, while for $m = 1$ it becomes identical to the Arditi-Ginzburg ratio-dependent model that was presented in chapter 1:

$$\frac{aN}{1+ahN} \xleftarrow{m=0} \frac{\alpha N P^{-m}}{1+\alpha h N P^{-m}} \xrightarrow{m=1} \frac{\alpha N/P}{1+\alpha h N/P} \qquad (2.10)$$

As discussed in section 2.1.1, the mutual interference parameter m typically belongs to the interval [0,1]. Negative values would occur with cooperative predators while values greater than 1 would characterize overcompensating interference. Ignoring such atypical cases, prey dependence and ratio dependence appear at the opposite ends of a spectrum of conceivable functional responses. The useful feature of model (2.9)–(2.10) is that parameter m provides a quantification of the position of a given system along this spectrum, preserving the saturating nonlinearity of the response in all cases.

2.1.4 Application to Literature Data

Experimental studies of responses to prey density alone or to predator density alone cannot be used to apply models (2.6)–(2.9). Measurements of the number of prey attacked are needed for various values of prey and various values of predator densities. Arditi and Akçakaya (1990) found in the literature 15 such trivariate data sets (listed in the legend of figure 2.5). Typically, they were experimental studies in which the number of prey attacked N_a was observed in response to increasing prey abundances $N_1, N_2, \ldots$, with this response repeated at various predator abundances $P_1, P_2, \ldots$ In addition to predators, studies on parasitoids were included if the species were reported to discriminate between healthy and parasitized hosts, in which case perfect discrimination was assumed. Such parasitoids were treated like predators.

For each data set, m was estimated for two models. The models differ in the way that the searching efficiency a is estimated at each value of P. The first model is the Hassell-Varley model, which uses the traditional eq. (2.2), as generalized by Jones and Hassell (1988) for situations with multiple values of P. The second model is the Arditi-Akçakaya model, which rests on the use of eq. (2.6). It corrects the HV model by allowing for a nonzero handling time h. For each value of P, a nonlinear regression is performed on the functional response $(N_a/P$ observed for $N_1, N_2, \ldots)$ to obtain the parameters a and h. In agreement with the AA model, the handling time h is forced to be unique for all values of P. For each model, the interference parameter m is then obtained by a weighted linear regression of log a against log P. A benefit of this method is that nonlinear least squares regression takes advantage of all available data to calculate standard errors for the estimated parameters. Therefore, it provides the standard error of m and the statistic $t=|m-1|/SE(m)$ can test whether m is different from 1. As an example, figure 2.4 compares the application of these models to the data shown in figure 2.2. The estimate of m increases from 0.36 with the HV model to 0.87 with the AA model. The latter value is not statistically different from 1.

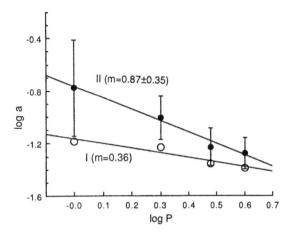

Figure 2.4. Application of both models to evaluate the searching efficiencies a and the mutual interference constant m in the data of Katz (1985). I: traditional HV model; II: corrected AA model. See Arditi and Akçakaya (1990) for more details. From Arditi and Akçakaya (1990), with kind permission of Springer Science+Business Media.

Figure 2.5 and table 2.1 show the application of the corrected AA model to the 16 data sets. In some cases, the errors of the searching efficiencies can be quite large, making the estimation of m rather uncertain. The errors of a are particularly large when the proportion of prey consumed is very low or very high. Good estimates of a are obtained when this proportion is between 0.2 and 0.8. Regarding the handling time, its estimate will only be good if the plateau of the functional response is experimentally attained. When devising experiments, the necessity of minimizing the estimation errors can serve as a guide for the choice of suitable values of T, P, and N.

Table 2.1 compares the values of m obtained with the two models. As expected, the AA model gives values that are consistently higher than those of the HV model based on a linear functional response. On average, m values increase from a typical 0.4 to a typical 0.75. This is evidence for the importance of nonlinearity in the functional response. If it were linear, both models would yield similar values of m.

All corrected values of m are significantly different from zero. This confirms the evidence of section 2.1.1, that is, the fact that the so-called parameter a of the Lotka-Volterra and Nicholson-Bailey models is, in fact, not a constant parameter. It should rather be considered a function of predator density, for example, varying as in eq. (2.4).

In the original analysis of Arditi and Akçakaya (1990), which has been reported here, the AA model was fitted in several steps: separate estimates of the searching efficiency a were obtained for the various values of P; then the interference parameter m was obtained by log-log regression. The advantage

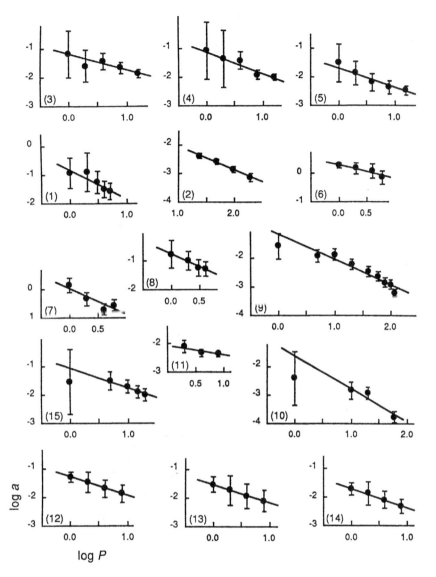

Figure 2.5. Application of the AA model to 15 data sets found in the literature. Values of a are plotted with one standard error. All graphs are to the same scale. See table 2.1 for the values of the slopes (−m). Cases (1)–(8) refer to predators and prey, and cases (9)–(15) to parasitoids and hosts. (1) Goldfish *Carassius auratus* and waterflea *Daphnia pulex* (Chant and Turnbull 1966); (2) flour beetle *Tribolium castaneum* adults and *Tribolium castaneum* larvae (Mertz and Davies 1968); (3)–(5) crustacean *Hyperoche medusarum* and herring larvae *Clupea harangus* (Westernhagen and Rosenthal 1976); (6) mites *Amblyseius degenerans* and *Tetranychus pacificus* (Eveleigh and Chant 1982); (7) mites *Phytoseiulus persimilis* and *Tetranychus pacificus* (Eveleigh and Chant 1982); (8) sea snail *Urosalphinx cinerea* and barnacle *Balanus balanoides* (Katz 1985); (9)–(10) wasp *Trichogramma evanescens* and eggs of grain moth *Sitotroga* (Edwards 1961); (11) wasp *Trichogramma pretiosum* and eggs of potato tuber moth *Phthorimaea operculella* (Kfir 1983); (12)–(14) wasp *Trioxys indicus* and cowpea aphid *Aphis craccivora* (Kumar and Tripathi 1985); (15) wasp *Trybliographa rapae* and cabbage root fly *Delia radicum* (Jones and Hassell 1988). From Arditi and Akçakaya (1990), with kind permission of Springer Science+Business Media.

Table 2.1. INTERFERENCE PARAMETERS M OBTAINED WITH THE HASSELL-VARLEY MODEL AND WITH THE ARDITI-AKÇAKAYA MODEL

Data Set	Original HV Model	Corrected AA Model		
	m	$m \pm SE$	$h \pm SE$	df
1	0.21	1.05 ± 0.36	6.8 ± 0.72 s	9
2	0.49	0.83 ± 0.09	4.6 ± 0.64 day	11
3	0.45	0.54 ± 0.16	2.1 ± 0.69 h	9
4	0.31	0.73 ± 0.17	5.3 ± 1.1 h	8
5	0.28	0.66 ± 0.17	4.1 ± 2.0 h	13
6	−0.05	0.50 ± 0.09	0.49 ± 0.04 h	9
7	0.41	0.92 ± 0.16	1.2 ± 0.22 h	14
8	0.36	0.87 ± 0.35	1.9 ± 1.3 h	11
9	0.55	0.89 ± 0.07	1.1 ± 0.16 h	65
10	0.50	1.14 ± 0.15	3.4 ± 2.0 h	12
11	0.11	0.33 ± 0.14	0.69 ± 0.073 h	8
12	0.58	0.64 ± 0.13	0.10 ± 0.07 min	4
13	0.58	0.62 ± 0.21	0.078 ± 0.15 min	4
14	0.58	0.66 ± 0.15	0.17 ± 0.09	4
15	0.39	0.70 ± 0.22	0.00 ± 4.4 h	119

The data sets are numbered as in the legend of figure 2.5. With the corrected AA model, the handling time h is estimated together with the interference parameter m. df is the residual number of degrees of freedom (number of observations minus total number of adjusted parameters). All values of m estimated with the AA model are significantly different from zero. After Arditi and Akçakaya (1990), with permission of Elsevier.

of this method is that the bias due to the HV model could be illustrated. An alternative method consists in performing a global nonlinear fit of the AA model to the whole trivariate data set (N, P, N_a). This fit yields simultaneously the three parameters α, h, and m. This direct method was generally used in more recent studies, such as those reported in the rest of this chapter.

Enriching the data set of Arditi and Akçakaya with seven additional studies, Skalski and Gilliam (2001) compared alternative functional response models to the AA model—which they called HV although they actually meant eq. (2.9) and not eq. (2.5). They used appropriate test statistics to account for the different numbers of parameters in the different models. Their online appendix B reports that the prey-dependent Holling model compared significantly better in one case only. Predator-dependent models other than AA fitted best in six cases. In all other cases, the AA model performed either better or equally well. Although the authors did not mention this explicitly, the AA model is the best choice for describing the whole data set with a single model. Unfortunately, the authors reported

[44] *How Species Interact*

the values of m for just five cases, of which only two were new, making it difficult to compare their findings with those of table 2.1.

A more recent meta-analysis by DeLong and Vasseur (2011) gathered 35 unbiased estimates of mutual interference (including all those of Arditi and Akçakaya and those of Skalski and Gilliam). They found evidence for a tendency toward ratio dependence, with a median m of 0.7 and a mean m of 0.8.

2.1.5 Does Interference Increase Gradually?

In section 1.6, we explained the gradual interference hypothesis that we have recently suggested. According to this hypothesis, interference should increase gradually from prey dependence (at low predator density) to ratio dependence (at high predator density). With this view, the two types of dependences are reconciled in one model and can appear in the same biological system. With the analytical tools used in this section, this means that the $\log a$ versus $\log P$ relationship should be curvilinear, starting with a zero slope ($m = 0$) at low P and bending downward to a strong negative slope ($m = 1$) at high P.

A visual inspection of figure 2.5 shows that at least four cases (cases 1, 9, 10, and 15) display this expected pattern. The other cases are linear in the range of P densities that were used but, of course, it is not impossible that higher and lower densities would show different slopes. In sum, the data, while not proving it, are consistent with the gradual interference hypothesis: the searching efficiency curves are either linear or, when nonlinear, are curved in the correct direction. We note that there remains a need for data on this topic and suggest that experiments with a very wide range of P densities be undertaken.

2.2 WASPS AND CHRYSOMELIDS: A FIELD EXPERIMENT

Direct measurements of functional responses in field conditions were performed by Schenk and Bacher (2002) and Schenk et al. (2005) with the paper wasp *Polistes dominulus* as predator and shield beetle larvae (*Cassida rubiginosa*) as prey. The 2002 study showed that the functional response was of type III. Since the predator density was not varied in that study, it was impossible to estimate interference. In the second experiment, Schenk et al. (2005) devised a way to manipulate the predator density as well. The predator density was either the natural density or augmented densities obtained by opening in the vicinity of the host plants one, two, or three cages containing paper wasp nests. Every day, the number of predators

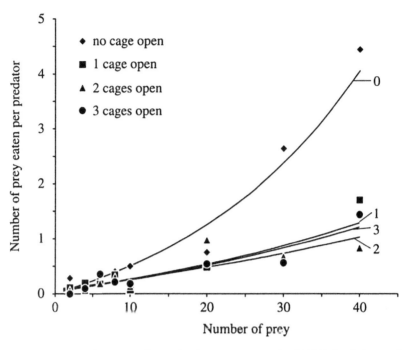

Figure 2.6. Functional responses observed with paper wasps preying on shield beetle larvae. The predator density was augmented by opening cages containing wasp nests in the vicinity of the experimental host plants. Opening cages clearly depressed the functional response, an indication of interference, but this effect did not change with the number of open cages. From Schenk et al. (2005), with permission of Wiley-Blackwell.

effectively hunting was determined by sweep netting during one hour (13h–14h). The prey density was set every morning by placing individual larvae in a patch of 200 thistle shoots, the natural host of the herbivorous larvae. Eight different prey densities were considered: $N = 2, 4, 6, 8, 10, 20, 30, 40$. The numbers of prey eaten each day were measured at 24 h intervals. The functional responses obtained with the different numbers of cages are shown in figure 2.6. Because they clearly appear to be convex, it is necessary to consider type III functional response models.

Since Schenk et al.'s original analysis made some questionable assumptions, we report here the major findings of a more rigorous analysis that identifies the model that best explains these data (Spataro et al., unpublished manuscript).

The direct observations during the 13h–14h activity hour show that the numbers of foraging predators did not differ significantly with 1, 2, and 3 open cages (see figure 1 of Schenk et al. [2005], where the number of active wasps even seems to decrease rather than increase with the number of open cages). Therefore, the only meaningful difference in predator density is between two values: the natural density (no cage open) and the augmented density (merging the cases of 1, 2, and 3 open cages). This merger is justified statistically by ANOVAs (see Spataro et al. [unpublished manuscript] for details).

The type III modification of the AA model is obtained by replacing the availability constant α by a function that increases linearly with N/P^m. This yields the model

$$g(N, P) = \frac{\left(\alpha + \beta \frac{N}{P^m}\right) \frac{N}{P^m}}{1 + \left(\alpha + \beta \frac{N}{P^m}\right) h \frac{N}{P^m}} \quad (2.11)$$

The standard type II AA model (eq. 2.9) is retrieved if β is set to zero. Thus, the type II model is nested into this type III model. Preliminary nonlinear fits of the model to the data and the graphical representation of the data (figure 2.6) all suggested that predator saturation did not occur within the range of prey densities used in the experiments. This makes estimation of the handling time mathematically impossible. Consequently, it is justified to set $h = 0$. This does not imply that the handling time is really zero but only expresses the fact that it is unnecessary to account for predator saturation. This simplifies the model to:

$$g(N, P) = \left(\alpha + \beta \frac{N}{P^m}\right) \frac{N}{P^m} \quad (2.12)$$

which is the type III variant of the Hassell-Varley model (2.5).

Likelihood-ratio tests can be used to compare nested models. A first comparison shows that the type III model (2.12) fits significantly better than the HV model, which is nested into it with $\beta = 0$ ($p < 0.04$). A second comparison shows that the simpler type III model with $\alpha = 0$, also nested into eq. (2.12), can be used with no significant loss of performance ($p = 0.2$). The same conclusions are reached with the information-theoretic criterion AICc (see Spataro et al. unpublished manuscript). Therefore, the model can be further simplified to

$$g(N, P) = \beta \left(\frac{N}{P^m}\right)^2 \quad (2.13)$$

with an estimate of the interference parameter of $m = 1.44$ and with a 95% CI of [0.98;2.19] obtained by bootstrap simulations. Interestingly, this value indicates overcompensating interference ($m > 1$) but, since the confidence interval includes 1, the parameter m is not significantly different from 1.

A further likelihood-ratio test can be applied to compare model (2.13) to the even simpler type III ratio-dependent model that is obtained by setting $m = 1$. This test yields a p value of 0.07, which is too high to justify the additional parameter.

In conclusion, the most parsimonious model describing the data is therefore quite simple:

$$g(N, P) = \beta \left(\frac{N}{P} \right)^2 \qquad (2.14)$$

This ratio-dependent model is the type III modification of the AG-DC model of table 1.1. Thus, the analysis of the elegant experiment of Schenk et al. (2005), which was performed in realistic field conditions, brings one more case in support of the ratio-dependent theory.

2.3 WOLVES AND MOOSE: FIELD OBSERVATIONS

It is hard to imagine a better example of a natural predator-prey system involving big mammals than wolves and moose in the Isle Royale National Park (Michigan). It provides an almost perfect "natural experiment," which is also the longest-running large mammal predator-prey study on earth. On this island in Lake Superior, 544 km² in size and located 24 km from the northern shore, wolves (*Canis lupus*) and moose (*Alces alces*) have been coexisting in a single-prey/single-predator system for over 60 years. Moose arrived at the beginning of the twentieth century while wolves arrived in the late 1940s. Detailed descriptions of field studies undertaken on this system are available in the many articles of R. O. Peterson and coworkers (e.g., see the references cited by Vucetich et al. [2002], or by Adams et al. [2011]). Immigration and emigration are nonexistent for both species. Moose have no other predators than wolves and they make up more than 90% of the wolves' diet. This makes this case unique since most other wolf populations are embedded in multiple-prey and multiple-predator food webs. Hunting is prohibited and these populations have effectively never been hunted. The population abundances have been monitored since 1959 with various methods, including aerial surveys since 1971. Moreover, counts of moose kills have been made during the winter aerial observations conducted since 1971. Thus, all the necessary data are available over several decades for making a direct assessment of the functional response: the prey abundances, the predator abundances, and the numbers of kills. This is truly exceptional in a long-term field study.

A first model comparison analysis was performed by Vucetich et al. (2002), using Akaike's information criterion. This analysis found that the ratio-dependent Arditi-Ginzburg model (1.14) performed significantly best of the 15 models considered. The robustness of this conclusion was later confirmed by Jost et al. (2005) with different mathematical and statistical methods. This second theoretical study also compared the results obtained at three different spatial scales, as explained below.

2.3.1 Wolf Social Structure and Spatial Scales

Wolves live in groups called packs. On Isle Royale, they form between two and five such packs every winter (typically three). From aerial surveys, the numbers of individuals in each pack are known. The aerial observations also make it possible to estimate the moose kills of each pack separately, by following the tracks in the snow. Each winter kill estimate was based on approximately 44 days of observations.

Vucetich et al. (2002) calculated the per capita kill rate as the number of kills made by a pack divided by the number of wolves in that pack, divided by the number of days during which that pack was observed. This was used as the dependent variable g of the functional response $g(N,P)$. The independent variables N and P were taken as the total moose and wolf numbers present on Isle Royale. The analysis was thus conducted on a mixed scale; that is, the dependent variable was measured on the pack scale while the independent variables were measured on the scale of the whole island. However, it may also be conceivable to consider an "island scale" only, with both dependent and independent variables measured on the whole island. Alternatively, one could work on the "pack scale," with wolf density, moose density, and the functional response all measured separately for each pack in its territory.

A problem with the island scale is that it ignores the pack structure altogether; it would doubtless be appropriate for predators hunting individually. The pack scale has a greater problem in that it ignores pack interactions and territory overlaps, and it divides the moose population into subpopulations supposedly confined to each wolf pack territory, which is not the case: the moose population remains a single population at all times. These considerations suggest that the mixed-scale approach followed by Vucetich et al. (2002) is sensible: it accounts for the wolf social structure and for the moose free movements over the island. The island scale would be reasonable because of its simplicity, for an aggregated description of the system. The least appropriate seems to be the pack scale. Nevertheless, there remained doubt as to which of the three spatial scales was the most biologically meaningful. For this reason, Jost et al. (2005) analyzed all three scales with the purpose of detecting general features of wolf predation that are independent of the choice being made.

Jost et al. analyzed the data from 1971 to 1998 (28 years). Accordingly, on the island scale, the data set is made up of 28 observations. On the pack scale, there are 85 different observations (because there are several packs every year). On the mixed scale, there are also 85 observations (85 values of the kill rate with only 28 different values of wolf and moose abundances).

2.3.2 Model Fitting and Model Selection Methods

The functional response g is by definition an instantaneous rate. In general, it can be denoted as $g(N,P,\theta)$, where $\theta=(\theta_1, \theta_2, \ldots)$ is the vector of parameters of the model. Given prey and predator densities, and observed kill rates g_{obs}, estimating the goodness of fit and the parameters for a particular form of g is a nonlinear regression problem that can be solved by minimizing the sum of squares:

$$SSE = \min_{\theta} \sum_{obs} \left(g_{obs} - g(N,P,\theta) \right)^2 \tag{2.15}$$

A potential issue is that actual observations are never instantaneous and that, during the observation time, the prey abundance declines because of predation. Thus, the observed kill rate gobs is not an instantaneous rate, as it should be. Therefore, the theoretical model that should be fitted through the data must be its integrated form, as was explained in previous examples [the quantity $N_a/(PT)$ defined in section 2.1]. In general, this leads to an implicit expression for N_a, which cannot be entered into standard software packages (see section 2.1). These complicated computations can be avoided only if it can be assumed that prey abundance does not change during the observation period. On Isle Royale, the actual moose depletion during the observations was of the order of 1.6% and never exceeded 5%, a value that can be considered negligible compared with other errors. Both Vucetich et al. (2002) and Jost et al. (2005) compared the results obtained with the instantaneous and integrated forms of the functional response models. In all cases, they found that the results were unaffected. For this reason, and for simplicity, here we report only the results obtained with instantaneous models.

As already mentioned, Vucetich et al. (2002) considered 15 different functional response models (including one constant, four prey dependent, five ratio dependent, and five predator dependent). In order to better focus on the biological issues of ratio dependence and saturation, Jost et al. (2005) considered a smaller set of nine models. We present here the performances of seven selected models. The performances of the other models can be found in the original publications, but none brought any interesting insight. The seven models were all presented in chapter 1, particularly in table 1.1. They are as follows:

- Lotka-Volterra (LV)
- Donor control (AG-DC)
- Hassell-Varley (HV)
- Holling type II (Ho)
- Arditi-Ginzburg (AG)
- Arditi-Akçakaya (AA)
- DeAngelis-Beddington (DAB)

Two are prey dependent (LV, Ho), two are ratio dependent (AG-DC, AG), and three are predator dependent (HV, AA, DAB). The first three in the list (LV, AG-DC, HV) do not display predator saturation while the other four (Ho, AG, AA, DAB) feature hyperbolic saturation. (Note that Jost et al.'s nomenclature of the models was different from the one used here.)

For model comparison, Vucetich et al. (2002) used Akaike's information criterion. However, this is more appropriate for selecting predictive models than for hypothesis testing, which is the present purpose (Burnham and Anderson 1998, 132). Jost et al. (2005) compared the models by model-based nonparametric resampling (Davison and Hinkley 1997). Models are compared pairwise to test the null hypothesis (H_0) that the worse-fitting model is the correct one and that the better-fitting model does so only by chance. To do this, 1,000 bootstrap samples were created by taking the worse-fitting model and randomizing the residuals (with replacement). Then, each pair of models were fitted again to the bootstrap samples. This method estimates the empirical distribution of the difference in goodness of fit between the two models under H_0, which can then be compared with the original difference. The proportion of bootstrapped differences that are greater than the original difference is an estimate of the smallest p value for which the difference is significant. H_0 is rejected if $p < 0.05$. The standard errors of the fitted parameters are estimated by standard nonparametric bootstrapping with 1,000 samples (Efron and Tibshirani 1993, 47).

2.3.3 The Wolf-Moose Functional Response Is Ratio Dependent

Table 2.2 compares the performances of the seven models. The columns labeled with SSE show the sum of squared errors (i.e., the criterion minimized by regression). The ratio-dependent Arditi-Ginzburg model fits best on the island scale and on the mixed scale. It shows the lowest values of SSE, and this value is not even lowered by the models AA and DAB, which have

Table 2.2. COMPARISON OF FITS OF SEVEN FUNCTIONAL RESPONSE MODELS G(N,P) TO THE ISLE ROYALE WOLF-MOOSE FIELD DATA

Model	Expression for g(N,P)	Dependence	Saturation	No. of Parameters	SSE Pack Scale	SSE Mixed Scale	SSE Island Scale
LV	aN	Prey	No	1	0.0402	0.0165	0.00176
AG-DC	$\dfrac{\alpha N}{P}$	Ratio	No	1	0.0724	0.0191	0.00188
HV	$\dfrac{\alpha N}{P^m}$	Predator	No	2	0.0402	0.0155	0.00135
Ho	$\dfrac{aN}{1+ahN}$	Prey	Yes	2	0.0170	0.0152	0.00175
AG	$\dfrac{\alpha N}{P+\alpha hN}$	Ratio	Yes	2	0.0132	**0.0116**	**0.00101**
AA	$\dfrac{\alpha N}{P^m+\alpha hN}$	Predator	Yes	3	**0.0124**	0.0116	0.00101
DAB	$\dfrac{aN}{1+ahN+cP}$	Predator	Yes	3	0.0132	0.0116	0.00101

At each scale, the best-fitting model is indicated with bold SSE. Overall, the best selection is the ratio-dependent AG model, which has two parameters. Note that, despite their additional free parameter, the AA and DAB models do not improve the fit on the mixed scale and on the island scale. See text for further discussion.

one more parameter than AG and should theoretically be able to fit better. On the pack scale, the procedure selects the model AA with a very high value of the interference parameter $m = 1.85 \pm 0.32$, that is, with overcompensation.

Note that, on all three spatial scales, the three models without saturation perform quite poorly. This proves that, biologically, the wolf kill rate is limited by some upper value h^{-1}. The value of h is quite independent of the scale with an estimate of 16–24 days, which seems a plausible value for handling one moose prey, although somewhat high (Thurber and Peterson 1993). Figure 2.7 illustrates how the model fits on the three spatial scales.

As explained in section 2.1.4, the AA model contains both the Ho and AG models as special cases for $m = 0$ and $m = 1$ respectively. For this reason, model selection can also be done by constructing confidence intervals of the parameter m and seeing if they include the values $m = 0$ or $m = 1$. This approach is followed in other sections of this book (e.g., sections 2.1, 2.2, 2.4, and chapter 3). As the AA model leads to overfitting on the mixed scale and on the island scale, we have preferred to base model selection on the goodness of fit (Jost and Ellner 2000; Jost and Arditi 2001). Nevertheless, it is of interest to study the confidence intervals of the estimates of m. Figure 2.8 shows the bootstrapped distributions of this parameter on the three scales:

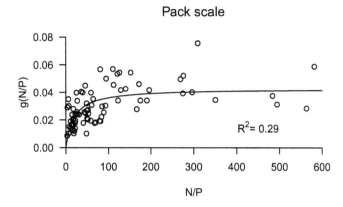

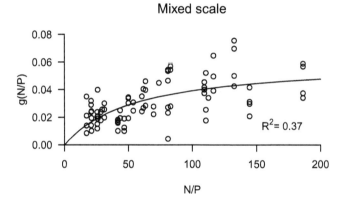

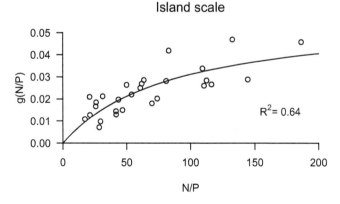

Figure 2.7. Fits of the ratio-dependent Arditi-Ginzburg model to the wolf-moose data on three spatial scales. R^2 is the coefficient of determination (proportion of variance explained). Note that the abscissa scale is not the same in all three graphs.

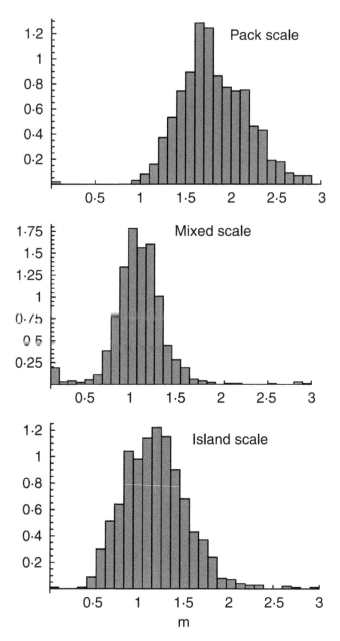

Figure 2.8. Distributions of the mutual interference parameter m of the Arditi-Akçakaya model, computed with a nonparametric bootstrap, on the three spatial scales. From Jost et al. (2005), with permission of Wiley-Blackwell.

they all show rather wide distributions, all very different from $m = 0$, while $m = 1$ seems a reasonable approximation for the island and mixed scales. On the pack scale, m is significantly greater than 1, indicating overcompensating interference: the whole wolf pack captures fewer moose as the pack size increases. Such high values of m have occasionally been reported in field studies of various systems (Hassell 1978, 2000; Sutherland 1996; Ponsard et al. 2000; Jost and Arditi 2001). We do not want to speculate here about possible reasons for this effect, both because no biological mechanism has been demonstrated and because it only appears in the pack scale analysis, which is the least reasonable approach.

Irrespective of the scale of analysis, the major result is the evidence of strong predator dependence. The underlying biological reasons need further field study. It will be difficult, if not impossible, to disentangle all mechanisms that are at work: social structure within packs, competition within packs, territoriality and competition between packs, partial prey consumption, and so on. Because many mechanisms are at work, fitting models that include only one mechanism at a time will only tell whether a particular mechanism can help explain the observed pattern, but reveals little about its importance compared with the other additional mechanisms. For this reason, it is appropriate to use emergent models such as AG and AA, which give a good phenomenological description of the global phenomenon. (Section 5.5 explains exactly what we mean by *emergent*.)

2.4 ADDITIONAL DIRECT TESTS OF RATIO DEPENDENCE

In the previous sections of this chapter, we reported a number of cases in which the influence of predator density on the functional response had been studied experimentally, with variations of both prey and predator abundances. In all these cases, the conclusion of a previously published analysis was, in general, that predators presented some degree of interference. We then initiated a reanalysis because we spotted potential bias. These reanalyses were done either on literature data (section 2.1) or in collaboration with the original authors (sections 2.2 and 2.3). In most cases, we established evidence for ratio dependence and in some cases for intermediate interference. We absolutely never could find any case of prey dependence.

An early illustration of ratio dependence was published by Bernstein (1981) using an acarine predator-prey system in a complex environment: when the numbers of prey and predators were varied with a constant ratio of 4:1, the number of prey eaten per predator did not vary significantly. However, this study did not explore the response to a full range of ratios. In this section, we report all the studies that, to our knowledge, were undertaken with the explicit intent of assessing the effect of predator density on the functional response in order to test the ratio-dependent hypothesis. They all report direct observations of the number of prey captured for various numbers of prey and various numbers of predators. Although the interaction times are not exactly of the same magnitude as the predator generation times, they are still long enough to be relatively exempt from the "instantist" misconception (see sections 1.3 and 5.3).

2.4.1 Bark Beetles

In a perfectly designed experiment, Reeve (1997) determined the nature of the functional response of the clerid beetle *Thanasimus dubius* attacking the bark beetle *Dendroctonus frontalis* under conditions close to natural. This bark beetle can mass attack and kill trees. When arriving on the tree, the xylophages can be preyed upon by the predators, but they become invulnerable once they have bored into the tree. In the experiments, pine bolts were used as arenas with three prey densities (100, 200, and 400 adults) crossed with three predator densities (10, 20, and 40 adults). After interacting for one day, the numbers of prey eaten were estimated from the remaining detached elytra (which are not consumed).

Reeve (1997) fitted to these observations the Arditi-Akçakaya model and the Arditi-Ginzburg model, using the implicit expressions of the integrated form of the models (as originally explained in Arditi and Akçakaya [1990]; see section 2.1 of this book). Both models adjusted equally well, explaining 73% of the variance (see figure 2.9). The estimate of the interference coefficient m in the AA model was 1.00 with a 95% confidence interval [0.58, 1.43]. Thus, prey dependence can be rejected outright because this interval does not include 0. With the best estimate $m = 1$, the AA model becomes identical to the ratio-dependent AG model. This study is remarkable because the author was aware of the mathematical problems

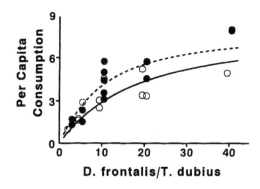

Figure 2.9. Fitted curves of the Arditi-Ginzburg ratio-dependent model to the data of a laboratory experiment on bark beetles. The variable on the abscissa is the prey/predator ratio. Separate curves are drawn for two independent experimental settings (dashed curve through black dots; solid curve through white dots). From Reeve (1997), with kind permission of Springer Science+Business Media.

of fitting functional responses and was able to devise an experimental design ideally tailored to test the ratio-dependent hypothesis.

2.4.2 Shrimp

Hansson et al. (2001) reported a set of four experiments undertaken with the zooplanktivorous shrimp *Mysis mixta*. The experiments differed depending on the prey species being used. One experiment used *Daphnia*, another experiment *Artemia*, and two experiments used mixed Baltic Sea zooplankton. Unfortunately, these experiments are somewhat imperfect and the mathematical analysis is likely to be biased. Although the interaction times were relatively long (several hours), the authors used the instantaneous forms of the functional response models instead of the integrated forms. However, the instantaneous forms remain approximately correct because the prey depletion was not severe. More critically, the authors assumed linearly increasing responses (i.e., they ignored predator saturation at high prey density). As explained in section 2.1, this assumption leads to underestimation of the interference coefficient m.

In the *Daphnia* experiment, the prey density was not varied while the predators were placed at four different densities. This univariate design makes it impossible to correct for the underestimation of m. Still, the observed searching efficiency declined hyperbolically with the predator density, indicating predator interference. The authors reported a value of $m = 0.53$. As this is an underestimate, the true value is unknown but, for certain, it is far from $m = 0$. In the *Artemia* experiment, the design was again univariate, with a single prey density and only two predator densities, and the (underestimated) interference was $m = 0.75$, again far from $m = 0$. In the two mixed zooplankton experiments, the design was bivariate (three prey densities crossed with two predator densities and interacting for 24 h) but, because the authors assumed a linear response, they used eq. (2.1) instead of eq. (2.6) and obtained the underestimated values $m = 0.63$ and $m = 0.66$. In conclusion, although the hypothesis $m = 1$ (corresponding to ratio dependence) cannot be tested because of the underestimation bias, the hypothesis $m = 0$ (corresponding to traditional prey dependence) can definitely be rejected.

2.4.3 Egg Parasitoids

Mills and Lacan (2004) worked with parasitoids and hosts rather than predators and prey: adult females of *Trichogramma minutum* attacking eggs of *Ephestia kühniella*. The advantage of such a system is that, contrary to prey, which disappear when eaten, hosts attacked by parasitoids remain present and can easily be recognized after 5–6 days from their dark color. In a pair of experiments, the functional responses of a single parasitoid and of three parasitoids were measured. From the visual inspection of the responses, the authors discarded the higher host densities in order to be certain that the parasitoids were maintained in the conditions of a linear functional response. In these conditions, eq. (2.1) is correct and the searching efficiency a can be estimated by nonlinear regression. In the first functional response series, a single parasitoid ($P = 1$) was present and this gave an estimate of $a = 1.32$. In the second series, three parasitoids were present simultaneously ($P = 3$) and the searching efficiency was reduced to $a = 0.37$, slightly less than one-third of the value in the first series. This supports the ratio-dependent hypothesis, which predicts that the product aP should remain constant.

In another experiment, a fixed number of hosts $N = 60$ was offered to varying numbers of parasitoids. Ensuring again that the parasitoids were in

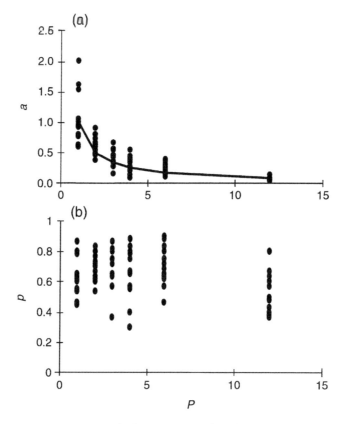

Figure 2.10. An experiment with the egg parasitoid *Trichogramma minutum*, displaying agreement with ratio dependence. (a) In relation to parasitoid density P, the searching efficiency a declines inversely proportionally. (b) The proportion of hosts attacked p remains constant. From Mills and Lacan (2004), with permission of Wiley-Blackwell.

the conditions of a linear functional response, this experiment showed that the searching efficiency a declined in inverse proportion with parasitoid density, and that the proportion of hosts parasitized was independent of parasitoid density, both relationships being in perfect agreement with the ratio-dependent hypothesis (figure 2.10). Note that these predictions are only valid because the functional response was linear. They would not hold exactly if the functional response were of the nonlinear saturating type.

2.4.4 Benthic Flatworms

Kratina et al. (2009) published the results of a well-designed microcosm experiment undertaken with the benthic flatworm *Stenostomum virginianum* preying on *Paramecium aurelia*. The numbers of prey eaten were measured from exhaustive counts of the prey remaining at the end of the interaction time. Compared to the experiments reported in sections 2.4.1 2.4.3, this experiment presents a number of improvements, particularly the following three:

1. Crossed with three prey densities (60, 160, 200 individuals per 900 µl) a very wide range of predator densities were applied (1–19 individuals per 900 µl).
2. Special consideration was given to low predator densities (1–5 individuals) because, as it was largely accepted that very high predator densities must have an effect on per capita predation rates, it remained uncertain whether this effect was also important at low densities (see our discussions in sections 1.5 and 2.1.5).
3. A relatively long interaction time was used (4 h), which is an intermediate time frame between instantaneous predation rate measurement and the worms' generation time (48–72 h).

Since the authors had determined in a previous experiment that this animal's functional response was sigmoid (type III), they developed modified forms of a number of functional response models (Ho, DAB, AA, and AG) so that they would become sigmoid. After correctly using the integrated forms of these models in nonlinear fittings, their four Akaike criteria were compared. Over the complete data set, the three predator-dependent models (AA3, DAB3, and AG3) outcompeted by far the prey-dependent model Ho3. The ratio-dependent AG3 fitted slightly less well than AA3 and DAB3, but this model has one less parameter. The interference parameter in the AA3 model was $m = 0.67 \pm 0.22$ (95% CI). It is particularly interesting that the same results were obtained when considering only the low range of predator densities (1–5 instead of the full range 1–19). In this case, the interference parameter in the AA3 model was $m = 0.66 \pm 0.45$ (95% CI). The major conclusion is therefore that, in this experimental system, there is evidence of strong interference, close to ratio dependence, even at low predator density.

2.5 IDENTIFYING THE FUNCTIONAL RESPONSE IN TIME SERIES

A number of time series exist in the literature for predator-prey systems in which the pair of abundances was periodically estimated over some length of time, but with no direct estimates of the number of prey eaten. That is, these data provide direct observations of predator-prey dynamics with no direct observations of the functional response. Thus, the question arises whether such data can be used to identify the dynamic model (including a specific functional response model) that best describes the observed dynamics. Jost and Arditi (2000) investigated the problem using artificial data, that is, data generated with a known model to which both process error and observation error were added. They found that with low to moderate noise levels, identification of the model that generated the data was generally possible. However, the noise levels that prevail in nature are stronger. Therefore, ecological time series observed in nature are not likely to provide reliable identifications of the functional response.

This was confirmed by the work of Jost and Arditi (2001) on real data. In relatively complex systems involving metazoa (mites in laboratory experiments and zooplankton in lakes), the results were inconclusive: either the selected model depended on the particular criterion being used, or no significant differences could be detected between the alternative models. In simple laboratory systems of protozoa, little predator dependence seemed to be present.

However, special attention was paid to the excellent laboratory work of Veilleux (1979) on the famous protozoan system *Didinium nasutum—Paramecium aurelia* (already studied by Gause [1934b, 1935a]). All simple models considered by Jost and Arditi (2001) showed poor agreement with these long cyclic time series. This indicated that more complex models were necessary and justified with such high-quality data. Using additional time series from Veilleux's master's thesis (1976), Jost and Ellner (2000) reconstructed the functional response nonparametrically, while also allowing for a delay in the reproductive functions of both prey and predator. Delayed effects not only improved the fit significantly but also revealed strong predator dependence in the reconstructed functional response. Two possible reasons why this helped are that the noise was averaged out and that the time intervals were brought to the time scale of population dynamics. The predator dependence was, in most cases, very well approximated by the ratio-dependent model (see further discussion in section 5.3).

2.6 CONCLUDING SUMMARY

In this chapter, we have attempted to assess the presence and the intensity of interference in all functional response data sets that we could gather in the literature. Each set must be trivariate, with estimates of the prey consumed at different values of prey density and different values of predator densities. Such data sets are not very abundant because most functional response experiments present in the literature are simply bivariate, with variations of the prey density only, often with a single predator individual, ignoring the fact that predator density can have an influence. This results from the usual presentation of functional responses in textbooks, which, following Holling (1959b), focus only on the influence of prey density.

Among the data sets that we analyzed, we did not find a single one in which the predator density did not have a significant effect. This is a powerful empirical argument against prey dependence. Most systems lie somewhere on the continuum between prey dependence ($m = 0$) and ratio dependence ($m = 1$). However, they do not appear to be equally distributed. The empirical evidence provided in this chapter suggests that they tend to accumulate closer to the ratio-dependent end than to the prey-dependent end. In several examples, we showed that the best estimate was exact ratio dependence ($m = 1$). It has been argued that interference observed in laboratory experiments is likely to be exaggerated by the confined conditions with unnaturally high densities (e.g., Hassell 2000). However, we have shown that the ratio-dependent AG model was also identified in all those rare field studies that could be analyzed (e.g., the wasps of section 2.2, the wolves of section 2.3, the bark beetles of section 2.4.1). In conclusion, it seems to be much more acceptable to assume ratio dependence as a null model (i.e., as a default unless proven otherwise) than to assume prey dependence as in the standard null model (Abrams and Ginzburg 2000).

CHAPTER 3

Indirect Evidence

Food Chain Equilibria

Equilibrium properties result from the balanced predator-prey equations and contain elements of the underlying dynamic model. For this reason, the response of equilibria to a change in model parameters can inform us about the structure of the underlying equations. To check the appropriateness of the ratio-dependent versus prey-dependent views, we consider the theoretical equilibrium consequences of the two contrasting assumptions and compare them with the evidence from nature. Thus, even without the large amount of direct measurements of kill rates indicating predator dependence, including ratio dependence (see chapter 2), a judgment about functional responses can be made based on qualitative predictions about population equilibria.

The major contrast in the predictions of the two theories is illustrated by table 3.1 (taken from Arditi and Ginzburg 1989). This table shows the responses of population equilibria in food chains of increasing length. For simplicity, we assume that primary production is represented by an exogenous input F to the lowermost level N, and that only the uppermost level suffers nonpredatory mortality. Consecutive levels are coupled by predator-prey terms, with predator production proportional to prey consumption. Mathematical proofs for the responses shown in table 3.1 are given in appendix 3.A.

According to the standard prey-dependent theory, in reference to the increase in primary production, the responses of the populations strongly depend on their level and on the total number of trophic levels. The last, top level always responds proportionally to F. The next to the last level always remains constant: it is insensitive to enrichment at the bottom because it is perfectly controlled by the last level. The first, primary

Table 3.1 RESPONSES OF FOOD CHAINS TO PRIMARY INPUT F

	Prey Dependence				Ratio Dependence			
Level	2	3	4	5	2	3	4	5
N*	→	↑	↓	↑	↗	↗	↗	↗
P$_1$*	↗	→	↑	↓	↗	↗	↗	↗
P$_2$*		↗	→	↑		↗	↗	↗
P$_3$*			↗	→			↗	↗
P$_4$*				↗				↗

Arrows show the variation of population equilibria to an increase of F, in food chains of lengths 2 to 5. Symbols: → no response; ↗ proportionate response; ↑ nonlinear increasing response; ↓ nonlinear decreasing response.

producer level increases if the chain length has an odd number of levels, but declines (or stays constant with a Lotka-Volterra model) in the case of an even number of levels. According to the ratio-dependent theory, all levels increase proportionally, independently of how many levels are present. The purpose of this chapter is to show that the second alternative is confirmed by natural data and that the strange predictions of the prey-dependent theory are unsupported. The arguments are based on comparisons between uncontroled ecosystems. Of course, they are liable to phenomena like community assembly, species invasions, and species losses, which may blur the picture. However, it will be seen that the conclusions obtained at this large scale agree with those obtained on a smaller scale in the controled experiments of chapter 2.

3.1 CASCADING RESPONSES TO HARVESTING AT THE TOP OF THE FOOD CHAIN

Ecological literature commonly refers to "top-down" and "bottom-up" controls. Sometimes these two concepts are put in opposition to each other, sometimes not. Sometimes they refer to short-term ecosystem reactions (so-called pulse perturbations *sensu* Bender et al. [1984]), sometimes to long-term equilibrium reactions. The latter ones were envisioned in the classical article by Hairston, Smith, and Slobodkin (1960), which is further discussed in section 3.4.

Ponsard et al. (2000) attempted to establish consistent definitions of the terms, but various people continue to mean many different things when they use the designations top-down and bottom-up. For this reason, we avoid using this terminology here. Instead, we refer to the trophic cascade in response to harvesting at the top of the food chain (this section), and to the response to enrichment at the bottom of the chain (next sections). In both cases, we concentrate on the long-term equilibrium responses. The reason for this is that the reaction of alternative theoretical constructs to pulse perturbations is the same (alternating signs of sequential trophic level responses; Bender et al. 1984). Thus, only the long-term effects can help us make a clear distinction between theories.

If top predators are eliminated or reduced in abundance, models predict that the sequential lower trophic levels must respond by changes of alternating signs. For example, in a three-level system of plants-herbivores-predators, the reduction of predators leads to the increase of herbivores and the consequential reduction in plant abundance. This response is commonly called the trophic cascade. In a four-level system, the bottom level will increase in response to harvesting at the top. These predicted responses are quite intuitive and are, in fact, true for both short-term and long-term responses, irrespective of the theory one employs (see appendix 3.B for the case of long-term responses of ratio-dependent chains).

A number of excellent recent reviews have summarized and meta-analyzed large amounts of data on trophic cascades in food chains (Shurin et al. 2002; Shurin and Seabloom 2005; Borer et al. 2005). In general, the cascading reaction is strongest in lakes, followed by marine systems, and weakest in terrestrial systems.

There are also significant exceptions, based on food webs, not simple food chains, with: (1) various species reacting to each other within a trophic level, not only across trophic levels; (2) omnivory connecting more than two trophic levels; (3) an undefined number of levels with, for example, one herbivore consuming the primary producing plants directly, and another one connected to the producer through intermediate trophic levels (Stibor et al. 2004; Gobler et al. 2005; Sommer and Sommer 2006).

With all the complexity of actual food webs (Cohen et al. 1990; Polis and Winemiller 1996; Caswell 2005; De Ruiter et al. 2005), it may be surprising to see that a concept based on a fixed number of trophic levels works so well and so commonly. We illustrate the trophic cascade with a wonderful diagram from Shurin et al. (2002) (figure 3.1). These authors summarized the results represented in this figure as follows:

> Predators reduced herbivore abundance in every system (i.e., the mean herbivore log ratio was always significantly less than zero. The herbivore response was greatest in lentic and marine benthos and weakest in streams and terrestrial systems. Predator effects ranged from a mean 17.3-fold reduction of herbivore density in lentic benthos to a 1.4-fold reduction in streams. Terrestrial food webs showed 1.6 times lower herbivore density with predators than without. Plant biomass often increased in the presence of predators; however, the magnitudes of the effects were generally smaller

than those on herbivores. The plant effect size ranged from a 4.7-fold increase in marine benthos to a (non-significant) 1.1-fold increase in terrestrial systems. The plant response was significantly greater than zero in four of six systems (every case except terrestrial systems and marine plankton). The systems where predators had the greatest effects on herbivores were the same as those where plants responded most strongly (lentic and marine benthos). Plants in the five aquatic systems collectively showed stronger responses to predators than those in terrestrial food webs. System type explained 28.6% of the variation in the plant log ratio, and 35.0% of the variation in the herbivore response. (Shurin et al. 2002, 786–787)

Any theory that claims to describe trophic chain equilibria has to produce such cascading when top predators are reduced or eliminated. It is well known that the standard prey-dependent theory supports this view of top-down cascading. It is not widely appreciated that top-down cascading is likewise a property of ratio-dependent trophic chains. We present this proof in appendix 3.B. Thus, at least qualitatively, the two theories cannot be distinguished by the evidence of cascading in response to harvesting at the top.

A more delicate issue is the quantitative difference in the strength of cascading in the two theories. This interesting question was raised by Robert Holt when reviewing a draft of this book. The absolute degree of cascading under ratio dependence depends on the degree of concavity of

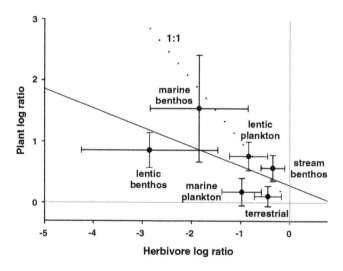

Figure 3.1. Comparison of the effects of carnivores on plants and on herbivores. The log ratio is $\log_e$ (abundance in presence of carnivores/abundance in absence of carnivores). Error bars are 95% confidence intervals. The effect of carnivores on herbivores is negative and significant in all six ecosystems. The effect of carnivores on plants is positive and significant in four ecosystems. The effects are weaker on plants than on herbivores (the regression slope b is $-1 < b < 0$). From Shurin et al. (2002), with permission of Wiley-Blackwell. See text for additional details.

the functional response (see appendix 3.B for details). In the linear part of it, that is, the leftmost part with prevailing donor control (figure 1.2), the prey abundance does not respond at all to partial changes of predator abundance (see section 1.4). However, the cascading response will be present if the change is sufficiently large to touch the high range (the rightmost part of figure 1.2), so that concavity of the functional response matters. Both theories agree in predicting strong cascading effects in response to total elimination of top predators. In a review, Estes et al. (2011) report a number of cases of aquatic and terrestrial ecosystems in which losses of apex consumers have had dramatic effects. These authors acknowledge that these effects are much clearer in cases of complete predator removal than in cases of partial changes.

In sum, strong cascading in response to complete removal of top predators is compatible with both theories and cannot discriminate between them. Weak cascading in response to partial changes at the top (or no response at all) would be in favor of the ratio-dependent view. Strong responses to partial changes at the top would argue in favor of the prey-dependent view only if predators are far from saturation. The distinction through cascading effects is therefore quite delicate. It is for equilibrial responses to enrichment at the bottom that predictions are strikingly different according to the two theories (see table 3.1). This is the subject of the next two sections.

3.2 ENRICHMENT RESPONSE WHEN THE NUMBER OF TROPHIC LEVELS IS FIXED

The goal of this section is to demonstrate that all levels of trophic chains in nature respond to enrichment at the bottom positively, and in most cases simply proportionally. This is not to say that the proportionality is the same for all levels—it is clearly different. Some may respond more strongly than others, but the point is that all do it in the same upward direction. Again, as in the previous section, we are addressing only long-term equilibrium responses. In this section, we assume that the number of trophic levels does not change with the degree of enrichment. We remind the reader that the predicted responses of alternative models are summarized in table 3.1. Whatever the chain length, the prey-dependent model predicts that the species that are 0, 2, 4 . . . levels from the top respond positively to enrichment, that those that are 1, 3, 5 . . . levels from the top do not respond or respond negatively; consequently, consecutive levels are never positively correlated. The ratio-dependent model predicts that all levels respond positively and, therefore, that they must all be correlated. The studies we summarize below and in the next section demonstrate that the predictions of prey-dependent models are in total contradiction with empirical findings from terrestrial and aquatic ecosystem studies. This review follows mostly Ginzburg and Akçakaya (1992).

A worldwide comparison of forest ecosystems of increasing productivity shows a responding increasing pattern both in plant and animal biomasses (Whittaker 1975, 224–226). Evidence supporting the ratio-dependent hypothesis is also provided by Ricklefs (1979, 623), who shows that wolf populations and their prey vary among localities in the same biomass ratio. These data were presented with figures in Arditi and Ginzburg (1989) and Arditi, Ginzburg, and Akçakaya (1991). McNaughton et al. (1989) compiled data on primary productivity, secondary productivity and consumption, and herbivore biomass from studies of terrestrial ecosystems (with typically three trophic levels), including desert, tundra, temperate grassland, temperate successional old field, unmanaged tropical grassland, temperate forest, tropical forest, salt marsh, and agricultural tropical grassland. The range of primary productivities covers three orders of magnitude. Their analysis shows that there is significant positive correlation between herbivore biomass and plant production.

In aquatic systems (often with four trophic levels), comparison of trophic biomasses across lakes has shown that both zooplankton biomass and fish biomass were positively correlated with primary productivity (and with phytoplankton density) (McCauley and Kalff 1981; Jones and Hoyer 1982; Hanson and Peters 1984; Pace 1984). In addition, all trophic levels are positively correlated with nutrient input as measured by phosphorus concentration (Deevey 1941; Yan and Strus 1980; Hanson and Leggett 1982; Jones and Hoyer 1982; Prepas and Trew 1983; Hanson and Peters 1984; Pace 1984; Stockner and Shortreed 1985; Persson et al. 1988). Reanalyzing the data of McCauley et al. (1988) on abundance of *Daphnia* and its algal food supply across 30 lakes, Arditi, Ginzburg, and Akçakaya (1991) showed that the positive relationship between the two levels did not deviate significantly from linearity (i.e., proportionality). A report comparing oceanic data along the coast of western North America shows highly significant monotonic relationships between phytoplankton, zooplankton, and fish abundances (Ware and Thomson 2005). Another study in an Australian estuary demonstrates very clear positive responses to enriched nitrogen at four trophic levels: phytoplankton, macrobenthic invertebrates, macrophagous fish, and piscivorous fish (Bishop et al. 2006). These patterns are not restricted to aquatic systems: a worldwide comparison of natural grasslands has also shown positive responses of plant and herbivore biomasses to increased productivity

(Chase et al. 2000). All these findings contradict the predictions of the prey-dependent models. They support ratio dependence, since correlations among the long-term averages of the biomasses of trophic levels across ecosystems are always positive.

Many ecosystem-level studies that concentrate on trophic interactions have been on lakes, probably because lakes provide well-defined, more or less closed systems with relatively distinct trophic levels. Some of these studies have been reviewed by Hanson and Peters (1984), Kerfoot and DeAngelis (1989), McQueen et al. (1986), McCauley et al. (1988), McCauley and Kalff (1981), Stockner and Shortreed (1985), and Carpenter (1988). Using these reviews as a starting point, Ginzburg and Akçakaya (1992) collated a large data set on 175 lake ecosystems, in which they examined the statistical relationships among different trophic levels. Since different studies used slightly different methods, most of the data were analyzed separately, except for the nutrient-phytoplankton relationship (see below). For each of these studies, a logarithmic regression was performed: $\log_{10} L_2 = a + b \log_{10} L_1$, where L_1 and L_2 are two consecutive trophic levels (including nutrients). The parameter b (the slope of the log-log regression) is an estimate of the reciprocal of the interference coefficient m between the two trophic levels. The case $m = 1$ corresponds to ratio dependence. Some of the statistical results reported by the original authors included repeated measurements from the same lake, so all the data sets were reanalyzed by first taking the average values for each lake and including each lake as a single data point. The original articles that were used in some studies could not be found. As a result, these data were not reanalyzed and the results were retained as reported by the authors (marked by an asterisk in table 3.2). The lakes of these studies are not included in the pooled sample for the nutrient-phytoplankton relationship (figure 3.2a). The regression analyses summarized in table 3.2 show that all but one of the slopes are significantly greater than 0; two-thirds of the slopes (10 out of the 15 for which the standard error either was calculated or was available) are not significantly different from 1; and one-third are significantly less than 1. These results demonstrate that all trophic levels respond in the same direction to an increase in productivity, and that the response is in most cases proportional.

In summary, one of the important differences of ratio-dependent theory from prey-dependent theory is the prediction of increases in the biomasses of all trophic levels in response to an increase in basal productivity. The evidence from both terrestrial and aquatic ecosystems and data on the trophic structure of lake ecosystems show that this expectation is generally correct. Moreover, the slopes of log-log relationships among trophic levels demonstrate that natural systems are closer to ratio

Table 3.2. REGRESSION ANALYSES FOR PAIRS OF TROPHIC LEVELS

Trophic Levels	$b\ (\pm SE)$	n	r^2	Reference	Figure
Nutrient vs.	1.335 ± 0.205	25	0.648	Jones and Hoyer (1982)	
phytoplankton	1.249 ± 0.133	49	0.669	Hanson and Peters (1984)	
	1.091 ± 0.093	12	0.932	Pace (1984)	
	1.061 ± 0.145	26	0.691	Prepas and Trew (1983)	
	1.013 ± 0.204	19	0.590	Stockner and Shortreed (1985)	
	0.997 ± 0.280	32	0.297	Deevey (1941)	
	0.884 ± 0.045	119	0.770	Ginzburg and Akçakaya (1992)	3.2a
Nutrient vs.	1.632 ± 1.371	11	0.136	Yan and Strus (1980)	
zooplankton	0.917 ± 0.083	49	0.723	Hanson and Peters (1984)	3.2b
	0.643 ± 0.084	12	0.855	Pace (1984)	
Phytoplankton vs.	0.719 ± 0.112	17	0.856	McCauley and Kalff (1981)*	
zooplankton	0.534 ± 0.067	49	0.572	Hanson and Peters (1984)	
	0.554 ± 0.084	12	0.812	Pace (1984)	3.2c
Nutrient vs. fish	1.566 ± 0.431	25	0.365	Jones and Hoyer (1982)	
	0.708	18	0.75	Hanson and Legett (1982)*	
Phytoplankton vs. fish	1.210 ± 0.137	25	0.77	Jones and Hoyer (1982)	3.2d

b is the slope of the log-log regression; n is the sample size; and r^2 is the variance explained by the regression. After Ginzburg and Akçakaya (1992), with permission of the Ecological Society of America.
*Values reported here as given by the authors.

dependence than to prey dependence. None of the systems shows the alternating positive, negative, and zero slopes predicted by the standard prey-dependent theory.

Occasionally, some field studies disagree with all theoretical predictions. Ponsard and Arditi (2000, 2001) used stable isotopes to establish that the litter-based food chain of a temperate deciduous forest consisted of two levels only: the macroinvertebrate detritivores and their predators. For two years, the equilibrium abundances were estimated along a natural gradient of annual litterfall (Ponsard et al. 2000). Detritivores were strongly correlated to local litterfall but predators hardly responded. This pattern, which contradicts both theories, suggests that the abundance of predators, much more mobile than detritivores, was averaged over the forest patches with varying amounts of litterfall.

It has been proposed (Gatto 1991; Gleeson 1994) that proportionality of equilibria could arise in a prey-dependent model from the predator mortality being a quadratic function of predator abundance (i.e., qP^2 instead of qP). This would represent a type of density-dependent mortality in the consumer population that would be independent of resource limitations or interactions with the food supply. This argument, seeking

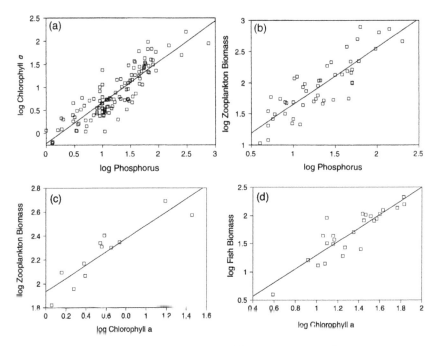

Figure 3.2. Relationships among lake trophic levels. (a) Nutrient concentration (total phosphorus in mg/m^3) vs. phytoplankton density (Chl. a in mg/m^3) for the database compiled by Ginzburg and Akçakaya (1992). (b) Nutrient concentration (total phosphorus in mg/m^3) vs. zooplankton biomass (mg/m^3) for the data set of Hanson and Peters (1984). (c) Phytoplankton density (Chl. a in mg/m^3) vs. zooplankton biomass (mg/m^3) for the data set of Pace (1984). (d) Phytoplankton density (Chl. a in mg/m^3) vs. fish biomass (kg/ha) for the data set of Jones and Hoyer (1982). From Ginzburg and Akçakaya (1992), with permission of the Ecological Society of America.

the cause of predator-prey correlations in non-resource-related mechanisms, appears weak in view of the universality of these correlations (Ginzburg and Akçakaya 1992; Akçakaya et al. 1995).

As noted above, the relationships among trophic equilibria are not always exactly proportional, indicating that the interference coefficients m are not always exactly equal to 1 (ratio dependence), although they definitely differ from 0 (prey dependence). This suggests that the more complicated models that are intermediate between prey and ratio dependence (e.g., those of table 1.1) may sometimes be more realistic. Since the biological mechanism for ratio dependence is interference, it is expected to be less pronounced in systems where consumer densities are low (see the gradual interference model of section 1.6). In addition, one may expect that the equilibria may not respond exactly proportionally when other limiting factors besides food come into play. It is therefore particularly interesting to see that, within the range of densities in natural ecosystems, the predictions of the ratio-dependent theory hold quite well, especially when compared to those of the prey-dependent models.

Our conclusion here is that, just as in the case of cascades in response to harvesting at the top, the preponderance of evidence indicates positive responses of trophic levels to enrichment at the bottom of the chain (often simply proportional).

3.3 ENRICHMENT RESPONSE WHEN THE NUMBER OF TROPHIC LEVELS INCREASES WITH ENRICHMENT

We deal here with the response of food chains to enrichment at the bottom when trophic levels are added with enrichment. A particular theoretical article has had considerable influence on ecological thought about enrichment responses: the article by Oksanen et al. (1981). This article expanded upon earlier work by Fretwell (1977) based on the prey-dependent theory. We focus on two theoretical predictions made in this article: (1) noticeable additions to the number of trophic levels as productivity increases, and (2) a curious behavior of the four-level system. We believe that both predictions emerging from this theory contradict the available data.

The idea that the number of trophic levels increases with basal enrichment is problematic. As proposed by Oksanen et al. (1981), new trophic levels are added at the top, and the primary producer population equilibrates at new abundances as enrichment at the bottom increases. At very low levels of enrichment, no consumer exists. As enrichment increases slightly, a primary consumer species can be sustained, but at abundances too low to support a secondary consumer species. Further trophic levels are added only as ecosystem enrichment occurs. While this unproven hypothesis is logically possible, it is at odds with available data: a large-scale review of a wide variety of ecosystems suggests that no relationship exists between primary productivity and the number of trophic levels (Briand and Cohen 1987; Cohen et al. 1990). A more recent analysis of data on temperate lakes (Vander Zanden et al. 1999) did find a positive correlation between primary production and food chain length, but Post et al. (2000) showed that the determining factor was in fact the lake size, not the primary productivity per se.

According to the Oksanen theory, enrichment can cause the abundance of particular trophic levels to either increase, remain constant, or decrease. The four-level system yields the greatest insight into this behavior. As shown in table 3.1 (fourth column), enrichment of a prey-dependent four-level system causes the top level (secondary carnivore) to increase in abundance, while the primary carnivore (third level) remains unchanged in abundance. Herbivores (second level) increase, while producers decrease in abundance. No intuitive explanation can account for this strange response of primary producers. Not surprisingly, most discussions are limited to the more palatable three-level system, which does not suffer from the prediction of decreasing primary producers with increasing enrichment.

Oksanen et al. (1981) discuss a single purported example of a four-trophic-level system that is consistent with their theoretical predictions. This example, from a study performed by Arruda (1979) on five farm ponds in Kansas, is claimed to display the expected responses for the four-trophic-level system (figure 3.3). Jensen and Ginzburg (2005) had a close look at this example and remained unconvinced by the results as they are presented. First, there appears to be an error in the transfer of Arruda's original results (figure 3.3, left panel) to Oksanen et al. (1981) (figure 3.3, right panel). In the third trophic level (primary carnivore), the data points are completely different: instead of showing the constancy in the third level expected by Oksanen et al.'s theory, Arruda's original figure actually displayed a significant decrease. C. X. J. Jensen contacted both J. A. Arruda and L. Oksanen asking for clarification, but neither of them could remember how this error crept into the highly influential 1981 article. The original Arruda article of 1979 had remained totally unnoticed.

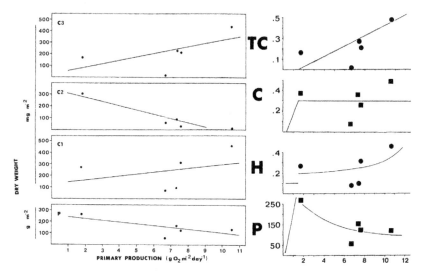

Figure 3.3. Relationships between productivity and equilibrial abundance in five farm ponds in order to test the predicted responses to enrichment in a prey-dependent four-level food chain (see table 3.1, fourth column). Left panels: in the original data (Arruda 1979), the only significant relationship is at level 3 (labeled C2); it contradicts both the prey-dependent and the ratio-dependent theoretical predictions. Right panels: the report of the same experiment by Oksanen et al. (1981) surprisingly shows different data at level 3 (labeled C). The curves were drawn to suggest that the responses of the four levels were consistent with the theoretical predictions, despite the nonsignificant nature of any attempt to fit a three-parameter curve to five data points. With permissions of the American Midland Naturalist (left) and of the University of Chicago Press (right).

Another concern arises when the significance of the curve fits is considered. All but one of the theoretically derived curves require at least three parameters to be depicted mathematically (one for the break point and two for the linear trend). In Oksanen et al. (1981), these curves are overlaid on the five Arruda data points (figure 3.3, right), giving the impression that the data are consistent with predictions. This impression is false, as it is impossible to reliably fit a three-parameter model to only five data points (Ginzburg and Jensen 2004).

A final worry concerns replication: if Arruda's results truly demonstrated that four-trophic-level systems were consistent with Oksanen et al.'s theoretical construct, we would expect other researchers to confirm them using a larger-scale approach more likely to yield statistically significant results. We know of no such attempt in over 30 years and therefore question whether Arruda's findings resulted from anything more than chance.

Accepting for a moment the idea that the number of trophic levels increases with primary productivity, the full response pattern predicted by Oksanen's theory can be summarized in two figures that show changes in pairs of trophic variables (Akçakaya et al. 1995). Figure 3.4a (based on Kerfoot and DeAngelis [1989]) shows the changes in the biomass of the first trophic level as the potential primary productivity increases. Figure 3.4b (based on Oksanen et al. [1981]) shows the changes in the biomasses of the first (producer) and second (herbivore) levels as a result of an increase in potential productivity (which is not shown). Sharp changes in the relationships arise as new trophic levels are added. In figures 3.4a and 3.4b, the last segments (labeled "4 levels") are drawn in a different style to emphasize that they correspond to the data shown in figures 3.4c and 3.4d, respectively. Ignoring the sharp changes and the peculiarity of the four-level segments, a fuzzy glance at each of the theoretical figures 3.4a and 3.4b suggests an approximate positive correlation between the two axes. Thus, the prediction of an increase in biomasses is contingent upon simultaneous changes in the number of levels. The problem is that such changes are not commonly observed, as explained above.

With a fixed number of trophic levels, the four-level case is most informative. In this regard, data from a very interesting study by Mills and Schiavone (1982) are especially important. We follow here Akçakaya et al. (1995), where these data were reanalyzed. Mills and Schiavone (1982) compiled abundance data on a series of 13 lakes. This study differed from others in its detailed identification of species in each of the lakes, which made it possible to ensure that each lake had four trophic levels. In these lakes, zooplankton density increased as a function of phytoplankton density (figure 3.4d), which in turn increased as a function of the nutrient

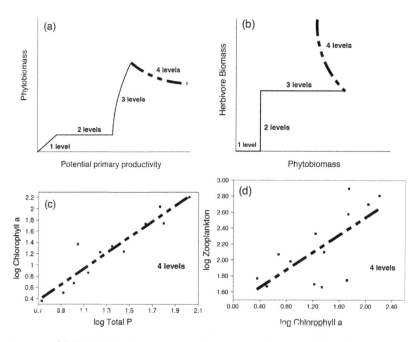

Figure 3.4. (a) Relationship between potential primary productivity and phytobiomass in the prey-dependent Oksanen theory (the different segments correspond to the assumed increase in the number of trophic levels). (b) Relationship between herbivore biomass and phytobiomass in the same theory. (c–d) Data dealing with four-level lakes (from Mills and Schiavone 1982), which must be compared with the last portion of the theoretical predictions (a–b). The data disagree with these predictions but agree significantly with the ratio-dependent model (which predicts simple positive correlations). From Akçakaya et al. (1995), with permission of the Ecological Society of America.

concentration (total phosphorus; figure 3.4c). Log-log regressions for both relationships give positive slopes that are significantly greater than zero. Since all of the lakes had four trophic levels, and an increase in the number of trophic levels is impossible, the prey-dependent theory would predict a decreasing function for both relationships (the last segments of the relationships in figures 3.4a and 3.4b), whereas the ratio-dependent theory correctly predicts the increasing relationships.

In the absence of supporting field data, we turn to the results of laboratory microcosm work, much of which claims to produce results consistent with Oksanen et al.'s (1981) predictions. As shown in table 3.1, the prey-dependent predictions suggest that in a simple two-trophic-level system, enrichment should cause no change in prey abundance and an increase in the predator population. These predictions were tested in two experiments by Bohannan and Lenski (1997) and by Kaunzinger and Morin (1998). What these articles show are significant increases in both predator and prey abundance (with predators responding more strongly than prey), a result that is inconsistent with the theoretical predictions. The most compelling empirical argument in favor of Oksanen et al.'s theory comes from the second experiment of Kaunzinger and Morin (1998). In this three-trophic-level system, they showed that the top predator disappeared consistently from the microcosms with the lowest nutrient inputs. To our knowledge, this experiment remains the only evidence for this prediction of Oksanen et al.'s theory. While we find this interesting, we worry that it could arise from a rather large departure from the assumptions of standard predator-prey theory. Like all batch culture systems, the microcosms of Kaunzinger and Morin violate the assumption of continuous nutrient input. For practical reasons, the microcosms were refreshed every four days. Given that in this four-day span one can get nearly 50 generations of the bacteria that occupy the producer level, we are concerned that starvation between nutrient pulses may be the reason for the elimination of the top predator from the microcosms with the lowest nutrient inputs. The approach of Kaunzinger and Morin was ingenious; we would like to see it replicated with a more frequent (i.e., closer to continuous) input of nutrients.

In sum, evidence from a variety of ecosystems suggests that comparable communities, varying in nutrient input level, differ only in the overall abundances at each trophic level (see section 3.2) and show none of the paradoxical behaviors predicted by the prey-dependent theory. Thus, both intuition and evidence suggest that enrichment causes increases in the abundance of all trophic levels.

A more recent article (Elmhagen et al. 2010) analyzes mammal population data in Finland. It considers the responses of trophic levels to primary productivity and top-down cascading simultaneously. The essential result is contained in the following quotation:

> The parsimonious model thereby revealed a top-down causal pathway where lynx biomass had a significant negative impact on fox biomass, whilst fox biomass had a significant negative impact on hare biomass. Although productivity had a significant positive impact on the biomass of all species, top-down control implied that fox biomass was lower than expected from productivity where lynx biomass was high, whilst hare biomass was lower than expected from productivity where fox biomass was high. The impact of top-down control was mediated by ecosystem productivity and stronger at low productivities.

The observed responses coincide precisely with our theoretical prediction. Apparently, these authors were not aware of our work, so this evidence is as independent as anything one can find in support of our views.

3.4 WHY THE WORLD IS GREEN

The very influential article by Hairston, Smith, and Slobodkin (1960, known as HSS) has driven interest in food chain research for 50 years. Its main role was to direct attention to equilibrial properties of trophic chains, not to short-term reactions, and thus to the so-called press rather than pulse experiments (Bender et al. 1984). The main message of the article was: the world is green because birds control herbivorous insects and, thus, prevent them from consuming plants excessively. So green material is left over every fall, which would have been totally consumed if not for the birds.

As we have attempted to explain in sections 3.1 to 3.3, the true picture is partially in agreement and partially in strong disagreement with this famous theory. The agreement comes when we consider the effect of top predator removal or reduction. Our ratio-dependent theory produces sign-altering effects (so-called cascading) down the food chain, just like the standard prey-dependent theory. So, if bird abundance is reduced, an increase of herbivorous insects and a decline in vegetation will follow. Since both theories agree entirely with this result in qualitative terms, experiments manipulating the top of the chain cannot be used to distinguish between the two views on consumer-resource interaction.

The strong disagreement comes when considering the equilibrium response of food chains to changes at the level of primary production. The HSS theory corresponds, in fact, to the three-level prey-dependent chain shown in table 3.1. According to this view, increasing vegetation production will not change the herbivore abundance in the long run (equilibrially). Increased plant production leads to increased herbivore production, which is picked up by the top predators. The increased predator population increases the consumption of herbivores, which, as a result, are maintained at the same abundance level. This is the sense in which birds "control" insect abundance. All the evidence that we collected (as discussed in sections 3.2 and 3.3) is in contradiction with this suggestion. In fact, the evidence is that all trophic level abundances increase in response to increasing primary production. The strength of the increase differs, but it is a monotonic increase at all levels. This is precisely what our ratio-dependent view suggested in table 3.1.

The three-level HSS theory is a weak case for testing the alternative theories. The first and third levels increase with primary production in both theories. The second level stays unchanged in the prey-dependent view and increases in the ratio-dependent view, but possibly with a low rate. So the distinction must be made between a slow increase and constancy just at one trophic level. The four-level extension of HSS is much more informative and is rarely invoked by the proponents of the prey-dependent view. As we explained in earlier sections, in this case, the plant equilibrium biomass must decline with increased productivity of this same level (see figure 3.4a). This predicted decline in the equilibrium abundance of the basal level has, of course, never been observed. We consider this the strongest evidence against prey dependence. So the HSS story of the next-to-top level being regulated by the top level breaks down as soon as the four-level system is considered.

It is peculiar that the HSS article presented the argument verbally in 1960 but the argument follows directly from the prevailing theory at the time: the Lotka-Volterra predator-prey model. It would also follow identically from any prey-dependent generalization. So it is an interesting lesson how this simplistic theory, apparently rooted in common sense, led to an intricate story that may appear theory independent. But isn't the ratio-dependent prediction even closer to common sense? Simple monotonic increase of the whole ecological community in response to enrichment is probably what a layperson would guess—and it turns out to be true.

In conclusion, we must judge the very influential HSS article as partially correct (for the effect of top-level manipulation) but completely wrong for the equilibrium response to enrichment, for which it is mostly used. A more recent book, *Why Does the World Stay Green?* (White 2005), attributes the greenness to evolved plant defenses and not to trophic interactions. This seems to be the best current view on this issue. For the purposes of our book, we just wanted to point out our disagreement with the HSS classic theory, which is, in fact, based on prey-dependent assumptions.

3.5 THE PARADOX OF ENRICHMENT

The paradox presented by the Rosenzweig-MacArthur model, a standard textbook generalization of the Lotka-Volterra model (see section 1.2), states that when the prey carrying capacity of a stable predator-prey system is increased sufficiently, the system begins to cycle (Rosenzweig 1971). Mathematically, the resulting structure is a limit cycle. As prey carrying capacity is increased further, this cycle brings one or both populations closer and closer to extinction. If the prey species goes extinct first, predator extinction will follow; if the predator species goes extinct first, this trophic level is lost and the prey stabilizes at its carrying capacity. Cited over 700 times, Rosenzweig (1971) is upheld as a classic example of an ecological paradox. We report here the study of Jensen and Ginzburg (2005) who explain how, although unproven empirically, the paradox of enrichment has quickly achieved the status of an ecological axiom. This enthusiasm is epitomized by the manner in which the word *paradox* has come to be interpreted. For Rosenzweig, the paradox was that enrichment, intuitively perceived as beneficial, had the potential to destabilize an ecosystem. In more recent usage, ecologists speak of "resolving" or "finding a solution to" the paradox of enrichment (Jansen 1995; Genkai-Kato and Yamamura 1999; Petrovskii et al. 2004): the paradox has capsized to mean that actual systems do not behave as Rosenzweig's model predict they should.

We have reviewed the literature on experimental attempts to demonstrate the paradox of enrichment and have found a disturbingly small number of studies in favor of the phenomenon. We contend that the need for experimental verification of the phenomenon is far from exhausted. What evidence is there for the paradox of enrichment? Several commonly misinterpreted examples from nature, as well as several experiments, merit discussion.

A commonly suggested example is the process of lake eutrophication. Enrichment of aquatic systems does appear to increase the carrying capacity of algal producers, generating a bloom that covers the lake. This bloom deprives the lake bottom of light, increasing aerobic decomposition and lowering the oxygen content of the water. It is this reduction in dissolved oxygen, not dynamic destabilization, that can cause the subsequent loss of top predators. While eutrophication does proceed from enrichment, its results are not paradoxical. If oxygen availability limits growth, any change that further reduces oxygen availability is expected to harm the system, particularly at trophic levels occupied by consumers. This description of eutrophication bears no resemblance to the predator-prey phenomenon described by Rosenzweig. We are puzzled by the fact that many ecologists still believe that eutrophication and the paradox of enrichment are connected.

Since Rosenzweig (1971) proposed that increasing the carrying capacity of a prey species could destabilize a predator-prey pair, a number of experiments have attempted to test this prediction. The first empirical work was performed in the *Didinium-Paramecium* system by Luckinbill (1973) and Veilleux (1979). When Gause (1934b, 1935a) originally studied this system, he was not able to obtain coexistence: the prey was brought to extinction, followed by the predator. Both Luckinbill and Veilleux showed that the system could be modified to produce persistence of predator and prey. Two modifications prevented the predator from consuming all of the prey: (1) the interaction rate of predator and prey within the system was reduced by the addition of methyl-cellulose, which thickens the medium and presumably reduces the searching efficiency of the predator; and (2) the availability of prey food was reduced. This second condition is commonly interpreted as evidence for the "paradox of enrichment in reverse," the system going from instability to stability when nutritional inputs are reduced (Harrison 1995). More recently, an experiment performed in a Rotifer-Algae system by Fussmann et al. (2000) showed that predator extinction resulted from enrichment. Like the Luckinbill and Veilleux experiments, the results of this experiment showed that reducing nutrient input can bring the system from a region of consistent predator extinction to a region of coexistence; unlike those much earlier experiments, the work of Fussmann et al. did not demonstrate a region of consistent dual extinction. These experiments did make a coarse argument in favor of the paradox of enrichment; increasing nutrient inputs does seem to destabilize the system. A recent article by Cabrera (2011) presents another chemostat study in which she attempted to test the theory of Fussmann et al. (2000). This test rejected the paradoxical predictions, and the results came out in favor of the ratio dependent alternative. We also remind readers that the nonparametrical analysis of Veilleux's experiments by Jost and Ellner (2000) had shown agreement with ratio dependence (see section 2.5).

When ecologists have looked for evidence for the paradox of enrichment in natural and laboratory systems, they often find none and typically present arguments about why it was not observed (Walters et al. 1987; McCauley et al. 1988; Watson and McCauley 1988; Leibold 1989; McCauley and Murdoch 1990; Watson et al. 1992; Persson et al. 1993; Mazumder 1994). Obviously, negative results receive less attention than positive ones, but we are surprised by just how minimal the impact of these negative results has been.

Why do researchers fail to observe this paradox in most experimental or any natural systems? If it is assumed that the paradox of enrichment could exist, the logical conclusion is that most of the experimental or natural systems in which it has been sought are not sufficiently simple. In other words, experimental conditions did not meet the theoretical assumptions and this is why experiments failed to demonstrate the paradox. Alternatively, if it is assumed that the paradox of enrichment does not exist, the logical conclusion is that other models of trophic interactions, ones that do not produce destabilization under enrichment, describe nature better.

The vast amount of theoretical effort in this area has been directed at the former explanation, producing a rich body of work showing that destabilization under enrichment can in theory be eliminated by a number of complicating mechanisms. The list of these potential mechanisms is long and continues to grow. It includes edible and inedible algae (Phillips

1974; Leibold 1989; Kretzschmar et al. 1993; Genkai-Kato and Yamamura 1999); density dependence of the predator death rate parameter (Gatto 1991); refuges and immigration (Abrams and Roth 1994); vulnerable and invulnerable prey (Abrams and Walters 1996); spatial heterogeneity (Nisbet et al. 1998); life-history traits that allow consumers to buffer the effects of low prey densities (McCauley et al. 1999); inducible defenses (Vos et al. 2004); and coevolution (Mougi and Kishida 2009). Generally, these theoretical explanations for the absence of the paradox of enrichment remain untested; some may even be untestable. Those few attempts to empirically confirm some of these increased complexity hypotheses have been unsuccessful (Murdoch et al. 1998). Models that incorporate various levels of additional complexity are difficult to falsify. With this additional complexity (and the associated addition of parameters), the danger of overfitting increases (Ginzburg and Jensen 2004). Algal species do differ in their edibility; some prey may be more vulnerable than others; and spatial heterogeneity and/or refuges are bound to be present in most natural systems. But we wonder. Isn't the lack of evidence for the paradox of enrichment an indication that no system will satisfy the assumptions of Rosenzweig-MacArthur dynamics?

3.6 DONOR CONTROL AND STABILITY OF FOOD WEBS

The original interpretation of donor control was very narrow, restricted to cases when predators consume only dying or dead individuals and thus have no effect on the prey population (Pimm 1982, 16–17). Prey abundance has a one-sided effect on predators and this one-way effect was termed *donor control*. As we showed in section 1.4, the ratio-dependent predation view gives a new meaning to the donor-control models. At high consumer density, ratio dependence implies that, even if predators consume healthy reproductive individuals, the per capita consumption may decline with P sufficiently to compensate for the number of predators.

Consider the predation terms subtracted from the prey equation and added to the predator equation in this special case. As shown in chapter 1 (eqs. 1.19, 1.20), the variable P will be missing in the prey dynamic equation even though a constant mortality induced by predation will still be present. This has a far-reaching consequence in food webs (not just food chains) of any complexity. If there exists an equilibrium in an arbitrary web, the Jacobian matrix defining the stability of this equilibrium will always be triangular, with zeros on one side of the diagonal. A triangular matrix always implies stability. Pimm (1982, 72) was first to obtain this result, but he then abandoned the study of donor-control models. The reason for this abandonment was based on simple qualitative evidence: with his narrow definition of donor control, the elimination of consumers would have no effect on the abundance of the other species. This contradicts ample evidence to the contrary. The situation is much more favorable with our admittedly simple but more general ratio-dependent donor-control model (eqs. 1.19, 1.20). If predators are removed, the mortality imposed by predation will disappear, and thus there will be consequences for the abundances of the underlying prey species.

The (in)famous stability-complexity issue would then resolve easily in donor-control models: there is no connection between the two properties. Stability is always guaranteed given that the equilibrium exists. It is the existence of the equilibrium rather than the stability that would relate with the number of species (complexity). This would, in turn, depend on how many predation mortality sources a given prey can support. This looks like a simple and reasonable structure. As long as the total of all mortalities applied to a given prey species does not exceed its reproduction rate, the equilibrium is feasible.

A relevant example is the work of Tilman (1994) showing that species competing for space may form a stable coexisting configuration if the interactions are strongly asymmetric. It is the triangularity of the interaction matrix that leads to stable coexistence in highly diverse plant communities. Another good example is hierarchical feeding in animal behavior, which is a very common social regulation mechanism. Competition among individuals can be represented by a triangular matrix: the subordinate individuals do not respond to competitive pressure from dominant ones. Why are these structures so common? They are structurally stable. Consumer-resource relations based on donor control may be an analogous mechanism of trophic web stability. An issue is direct interference (i.e., not through the consumption of a common resource). If such interference is present, the matrix loses its triangular structure and all the above simplicity disappears. It is an open research question whether interspecific interference that is not related to resource consumption is significant in shaping multispecies communities. At least, our simple stable structure includes the effect of consuming common resources.

The stability of a triangular matrix is preserved if all elements are small instead of exactly zero on one side of the diagonal. So deviations from precise donor control are acceptable in the above framework as long as the influences remain strongly asymmetric: the predator dynamic equations must depend strongly on prey abundance and the prey dynamic equations must depend only weakly on predator abundance.

In order to assess the usefulness of the donor-control approximation, one must check whether the interaction strength is strongly asymmetric or close to symmetry, as traditionally assumed. This is not an easy question to answer since simultaneous data on predator-prey dynamics are rare, and most of the available ones are only for cyclic species. In the case of cyclic species, we have argued before (Inchausti and Ginzburg 1998), on the basis of the original argument by Bulmer (1975), that predators may not have much effect on the prey. They are just "riding" the food cycle, with the cause of the cycle located in the prey. We have reinforced this argument with various kinds of evidence in Ginzburg and Colyvan (2004). The single-species view is a plausible alternative to the predator-prey view as the cause of cycles (see section 5.4). Cyclic populations are, however, a minority. In order to check the potential centrality of donor control (in our generalized sense), one has to demonstrate the asymmetry on a variety of natural data, not necessarily cycling. The work we are proposing has not been done. Moreover, Jost and Arditi (2000, 2001) have shown that the noise levels that prevail in natural time series make it impossible to decide between alternative models for the underlying interaction. We would like to encourage further explorations of coupled time series. We believe that the answer may lie in selecting the appropriate time scale for evaluating the symmetry or asymmetry of the predator-prey interaction. On a longer time scale, the interactions can prove to be asymmetric because the noise may not hide it as much as on a short time scale. This is illustrated in section 5.3 with a reanalysis of the Isle Royale wolves by Jost et al. (2005).

Note that in donor-controlled systems, predator removal experiments would still work as they usually do: prey abundance would increase, being released from the imposed mortality. So it seems to us that there is no obvious contradiction in our simple view as long as we see donor control as a special case of ratio-dependent predation rather than as "consuming the dying."

Donor control is a quite radical and potentially far-reaching simplification. As we showed in section 1.4, it implies that variation in predator numbers (over some reasonable range, where prey densities are not so high that individual predators are satiated) might not alter all that much the total mortality imposed by the predator population upon the prey population. In controled experiments, variations in predator "dosage" should not influence the total prey mortality, as reflected in various metrics. There are many fewer such experiments than there should be. We would like to encourage such attempts, which will help to distinguish between the ratio-dependent and prey-dependent abstractions in a clear and practically important way.

CHAPTER 4

How Gradual Interference and Ratio Dependence Emerge

As we have mentioned several times in this book, already in Chapter 1, the traditional Lotka-Volterra interaction term aNP is nothing other than the law of mass action of chemistry. It assumes that predator and prey individuals encounter each other randomly in the same way that molecules interact in a chemical solution. Other prey-dependent models, like Holling's, derive from the same idea. Two conditions are required for the law of mass action to work correctly: the concentrations of the reacting molecules must be low and the solution must be well mixed. Therefore, an ecological system can only be described by such a model if conspecifics do not interfere with each other and if the system is sufficiently homogeneous, both with respect to the physical environment and with respect to individual spatial behavior. Bacteria at low concentration, feeding in a well-stirred vessel, are a good example, growing according to Monod's model. In heterogeneous situations, arising from any deviation from this ideal situation, we have argued that the predator-prey interaction should become predator dependent, and we have also argued that ratio dependence is likely to be a reasonable simpler approximation. For bacteria, this was shown in the 1950s by Contois (1959). In this chapter, we follow a mechanistic approach and we will demonstrate that spatial heterogeneity, be it in the form of a prey refuge or in the form of predator clusters, leads to emergence of gradual interference or of ratio dependence when the functional response is observed at the population level.

This will first be illustrated experimentally in aquatic microcosms. We will show that homogeneously distributed filter feeders follow the law of mass action, whereas the existence of clusters, either arising spontaneously or imposed artificially, lead to ratio-dependent consumption.

A simple two-patch model shows that donor control (i.e., an extreme form of ratio dependence) can result from one patch being a prey refuge and the other patch being subject to intense predation. The same fundamental mechanism functions in a more realistic model, in which the spatial distribution of both species is followed explicitly. Predator taxis rules lead to the formation of clusters that move constantly in space. In the predator clusters, the prey are consumed almost entirely by the predators while the areas outside the predator clusters function as prey refuges, constantly supplying new prey to the predator clusters. Consequently, the population-level functional response is proportional to the ratio of the total prey to the total predator population, even though mass action is assumed at the local scale. An interesting application of these models is the explanation of successful biological control of insect pests, which remains an unintelligible paradox if traditional prey-dependent models are used (see section 4.4.1).

Strict ratio dependence, as outlined above, is a frequent outcome but not necessarily universal. If the conditions that are required for the application of the mass action law are fulfilled, at least approximately, prey dependence can prevail. We present two mechanistic individual-based models that illustrate how, with gradually increasing predator density and gradually increasing predator clustering, interference can become gradually stronger. Thus, a given biological system, prey dependent at low predator density, can gradually become ratio dependent at high predator density.

4.1 EXPERIMENTAL EVIDENCE OF THE ROLE OF PREDATOR CLUSTERING ON THE FUNCTIONAL RESPONSE

In the introduction to chapter 2, we explained why direct measurements of kill rates are problematic. Correct observations must be done on the time scale of population dynamics, not on the time scale of individual behavior. By definition, population equilibria occur on the correct time scale and, as shown in sections 3.1–3.3, can give insights about the underlying interaction dynamics. The many examples, particularly those of section 3.2, indicate that equilibrium patterns observed in natural ecosystems disagree with the predictions of prey-dependent models and agree with those of ratio-dependent models. However, natural ecosystems present the obvious problem that they are quite complex; it could be argued that their properties are not necessarily determined by the sole predator-prey interaction. In this section, we present a series of experiments designed to discriminate the alternative models in predator-prey systems that are both simple and on the appropriate time scale, that is, with observations made at equilibrium. We will show that the prey-dependent model describes predator-prey dynamics in homogeneous situations, whereas clustered situations are better described by the ratio-dependent model.

4.1.1 An Aquatic Microcosm Experiment

The basic experimental setup was documented with the preliminary experiments of Arditi, Perrin, and Saïah (1991). Filter-feeding zooplanktonic cladocerans are reared in a flow-through system where water containing algal food is pumped into serially arranged containers from which consumers cannot escape (figure 4.1). Starting with an inoculum in each container, the cladoceran populations are allowed to increase or decrease until equilibrium is reached. This design permits discrimination of the two forms of functional response by observing the patterns of equilibria of consumer populations only. The calculations of the alternative model predictions are straightforward and can be found in the appendixes to Arditi, Perrin, and Saïah (1991). The prey-dependent hypothesis predicts that the first container should stabilize at some population P^* and that all subsequent containers should become extinct:

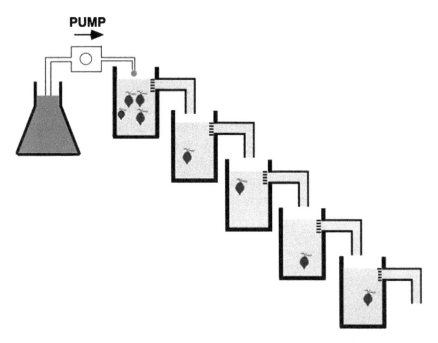

Figure 4.1. General experimental setup of the cascade of five chambers. Algae are pumped at constant flow from a stirred tank into the cascade. Starting with an inoculum of cladocerans in each chamber, populations increase or decline until an equilibrium is reached after a few weeks. Filters prevent the passage of newborn cladocerans through the cascade. After Arditi and Saïah (1992), with permission of the Ecological Society of America.

Prey-dependent pattern: $P^*, 0, 0, 0, \ldots$

This prediction is due to the fact that the prey equilibrium is set by the predator equation only (eq. 1.10, second row) and does not depend on prey "production" (i.e., inflow rate and concentration). Once the algal density has reached this critical value in the first container, the food concentration flowing into the following containers is insufficient to sustain a consumer population.

The ratio-dependent hypothesis predicts a very different pattern. Consumer equilibria decline in a geometrical sequence because the predator equation (eq. 1.10, second row) now sets the equilibrium *ratio* to a constant value:

Ratio-dependent pattern: $P^*, bP^*, b^2 P^*, b^3 P^*, \ldots$ (with $b<1$)

Full details on the experimental setup are given in Arditi and Saïah (1992), where additional results to those reported here can also be found. The prey consisted of live *Chlorella vulgaris*, which were introduced at a constant flow into the experimental cascades. We used two different species of cladocerans: *Daphnia magna* and *Simocephalus vetulus*. *Daphnia* swims constantly and is more or less homogeneously distributed in space. *Simocephalus*, on the other hand, has a special gland on the side of the head for attaching to substrates and, in the experiments, it generally attached to the walls of the containers; in nature, it rests on leaves of macrophytes. We expected that *Daphnia* would follow the prey-dependent model while *Simocephalus* would follow the ratio-dependent model. Single clones were used for each species and they reproduced parthenogenetically for the whole duration of the experiments. The experiments were run until reasonably stationary levels were reached in each container (in 4–7 weeks). Stationarity was checked in terms of number of individuals, in terms of biomass, and in terms of size structure. The equilibrium values were estimated as the average of the last 3 weeks.

4.1.2 Predator Aggregations Lead to Ratio Dependence

Daphnia maintained a stable population in the first container only (figure 4.2). In the other containers, the populations became progressively extinct starting with the last and ending with the second. *Simocephalus* presented completely different dynamics and patterns of equilibria, reaching stable populations in all chambers of the cascade (figure 4.3). This equilibrium pattern followed approximately a geometric decrease from the first to the last chambers. (However, the last container exhibited a significant upward deviation, which may be due to some food production in the upstream containers.) The results are consistent with the expectation that different spatial behaviors should lead to different mathematical models for the functional response: aggregated distributions being associated with ratio dependence and homogeneous distributions with prey dependence.[1]

This was corroborated by the results of additional experiments in which the spatial distribution was artificially modified. The basic idea was to force each species to follow the spatial distribution of the other. With *Daphnia*, we added in each container a central cylinder made of fine mesh fabric, permeable to algae, making 43% of the container inaccessible: the individuals were forced to swim and feed along the walls and the bottom of the containers. With *Simocephalus*, we modified the spatial environment by offering a large number of additional resting places all over the volume of each container. Thus, *Daphnia* was forced to a clustered distribution while *Simocephalus* was allowed to occupy a more homogeneous distribution. We predicted that these modifications would cause each species to follow the opposing model with respect to the unmodified situation.

With *Daphnia*, the addition of an inaccessible cylinder created a refuge for the prey. This modification changed the pattern of population equilibria along the cascade (figure 4.4). As in the unmodified *Simocephalus* cascade (figure 4.3), we obtained stable populations in all chambers with an approximate geometric decrease. (The lower-than-expected population in the first container may be due to the fact that the inflow of algae dropped from a lower height than in the following ones, badly homogenizing the distribution of algae.) In the modified experiment with *Simocephalus*, with increased spatial homogeneity, the populations ended with a nonzero equilibrium in the first container only (figure 4.5), that is, a pattern identical to the original *Daphnia* experiment (figure 4.2).

Throughout this book, we have suggested that the reasons that natural systems seem to respond in a ratio-dependent manner must lie in the factors that make real ecosystems impossible to reduce to chemical solutions—most of all, spatial and temporal heterogeneities. The present experiments fully support the assertion that homogeneous conditions lead to prey dependence while heterogeneous conditions can be characterized by ratio dependence.

---

1. Strictly speaking, the pattern of geometrically declining populations is also compatible with other predator-dependent functional responses such as the DeAngelis-Beddington model (Ruxton and Gurney 1992). Declines are also predicted by the Hassell-Varley and the Arditi-Akçakaya models. However, since the observations are equally well explained by the more parsimonious ratio-dependent model, there is no reason to prefer the other models, which contain one additional parameter (Arditi et al. 1992).

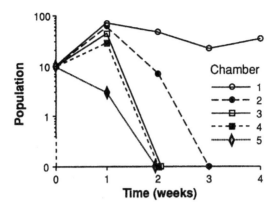

Figure 4.2. Population dynamics (log scale) of *Daphnia magna* in the five chambers of the cascade. With the exception of the first chamber, all populations declined to extinction, as predicted by the prey-dependent hypothesis. After Arditi and Saïah (1992), with permission of the Ecological Society of America.

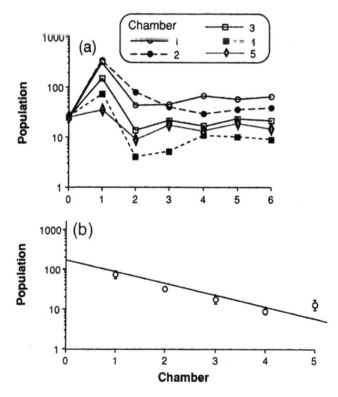

Figure 4.3. (a) Population dynamics (log scale) of *Simocephalus vetulus* in the five chambers of the cascade. All populations stabilized. (b) Population equilibria (log scale) in each chamber. Each point is the average of six values (last three weeks of two replicates). The equilibria follow approximately a geometric sequence, as predicted by the ratio-dependent hypothesis. After Arditi and Saïah (1992), with permission of the Ecological Society of America.

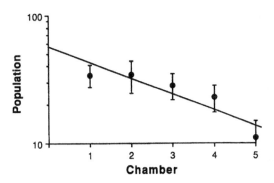

Figure 4.4. Population equilibria (log scale) of *Daphnia magna* in modified setup. Addition of a cylinder in the center of each chamber (permeable to algae but not to *Daphnia*) changes the pattern of equilibria to a geometric sequence, consistent with ratio dependence. After Arditi and Saïah (1992), with permission of the Ecological Society of America.

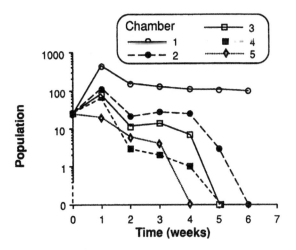

Figure 4.5. Population dynamics (log scale) of *Simocephalus vetulus* in modified setup. Addition of more resting places in each chamber led to extinction in the last four chambers, that is, to a pattern consistent with prey dependence. After Arditi and Saïah (1992), with permission of the Ecological Society of America.

4.2 REFUGES AND DONOR CONTROL

We present in this section a model (Poggiale et al. 1998) that is, in some way, an idealized mathematical representation of the experiments described in section 4.1. We represent in a purposely simplified way the two biological factors that we view as the fundamental causes of ratio dependence: (1) the existence of two distinct time scales, a fast behavioral time scale and a slow demographic time scale; and (2) the existence of spatial heterogeneity, represented by a prey refuge. We show that the dynamics of the predator-prey system on the demographic time scale become donor controled. Even though predators may control the prey density locally and on the behavioral time scale, the prey dynamics are independent of predator density globally and on the demographic time scale: the presence of predators has a dramatic depressing effect on prey abundance, but the prey dynamics are apparently independent of the predator dynamics. In other words, the prey population dynamics depend in a switchlike manner on the presence or absence of predators, not on their actual density.

In section 4.2.1, we present a rationale leading to a mathematical model that contains large and small terms in order to take into account the different time scales on which different processes take place. The feeding rate on the fast behavioral time scale is assumed to be prey dependent. We then prove that, on the slow time scale, the dynamics of predators depend on those of the prey and that the prey dynamics become apparently independent of predator dynamics. In section 4.2.2 we show how ratio dependence emerges in even more general conditions.

4.2.1 A Simple Exploratory Theoretical Model

We apply methods of perturbation theory, which enable us to show how the details of individual behavior on the fast time scale determine the population dynamics on the slow time scale. Predator clustering means that predation pressure varies from one place to another: there are places where predation is strong, places where it is weaker, and places where it is practically nonexistent. In the present model, this is simplified as two patches only: one refuge patch, inaccessible to predators, and one predation patch, each homogeneous (figure 4.6). As previously explained in this book, and as Arditi and Saïah (1992) demonstrated experimentally, one can expect prey-dependent predation in spatially homogeneous systems. Therefore, we assume that, in the predation patch, the functional response is the familiar Holling prey-dependent expression. Commonly, when predators are placed together with their prey in a homogeneous space, their dynamics bring the prey density to a very low value. As a result, the prey becomes extinct, followed closely by the predator. Such dynamics have frequently been observed in laboratory experiments (Gause 1934b, 1935a; Luckinbill 1973, 1974). In such a case, predation must be very intense. Therefore, we assume that (in patch 1) predation is very strong and much more important than all other processes. The refuge patch (patch 2), where the prey are free from predation, will make coexistence possible. We assume that the prey migrate between the two patches randomly. The prey growth in each patch is assumed to be logistic. The numerical response is taken to be proportional to the feeding rate. All these assumptions translate into the following equations:

$$\frac{dN_1}{d\tau} = -\frac{aN_1 P}{1+ahN_1} + \varepsilon\left[rN_1\left(1-\frac{N_1}{K_1}\right) + m_{12}N_2 - m_{21}N_1\right] \quad (4.1)$$

$$\frac{dN_2}{d\tau} = \varepsilon\left[rN_2\left(1-\frac{N_2}{K_2}\right) + m_{21}N_1 - m_{12}N_2\right] \quad (4.2)$$

$$\frac{dP}{d\tau} = \frac{eaN_1 P}{1+ahN_1} - \varepsilon qP \quad (4.3)$$

where m_{ij} is the migration rate from patch j to patch i, τ is the time, and the other symbols are as usual. Units are chosen such that all parameters (except ε) are of the order of 1. The small dimensionless parameter $\varepsilon \ll 1$ is used to indicate explicitly that predation is much stronger than the other processes. As we will show, the system (4.1–4.3) can have positive equilibria for both prey and predator densities.

If perturbation theory is applied (see Poggiale et al. [1998] for mathematical details), it can be shown that, on the slow time scale measured by $t = \varepsilon\tau$, the three-dimensional system (4.1–4.3) converges (for small ε) to the following two-dimensional system, with $N = N_1 + N_2$ denoting the total prey abundance:

$$\frac{dN}{dt} = rN\left(1-\frac{N}{K_2}\right) - m_{12}N \quad (4.4)$$

$$\frac{dP}{dt} = em_{12}N - qP \quad (4.5)$$

In eq. (4.4) we see that the prey dynamics are independent from predator abundance. The dynamics of predators (eq. 4.5) depend, however, on prey abundance. This is a characteristic feature of donor control: the predators have no dynamic effect on the prey, whereas they themselves depend on prey availability. On the detailed level of description, the behavioral, fast-time-scale feeding rate in the feeding patch was prey dependent. When we look, however, at the global environment and the slow time scale on the population level, donor control emerges.

However, this donor control has limits: if the predators are reduced, by external factors, to very small abundance or are even removed, the

Figure 4.6. A simple two-patch model. Predators cannot enter patch 2, which is a prey refuge. In patch 1, predation is a fast process. Reproduction, mortality, and migrations are slow processes.

assumption of strong predation no longer holds. In such a case, the prey population will increase to a value determined by the carrying capacities of both patches, $K_1 + K_2$, and may be much higher than in the presence of predators. Thus, an efficient predator can suppress the abundance of its prey by confining it to the refuge but, as less and less prey are available outside the refuge, predators have an ever-decreasing effect on the prey population. Donor-control dynamics described by eqs. (4.4, 4.5) rapidly begin to dominate. In this regime, life for the prey is simple: either stay in the refuge and reproduce, or leave the refuge and die as food to the predators.

Poggiale et al. (1998) provide numerical examples and simulations. They show that the above method gives an excellent approximation for $\varepsilon < 0.1$. In other words, donor control appears if predation (on the behavioral time scale) is about 10 times faster than reproduction (on the population dynamic time scale). Biologically, this is quite realistic. Additional simulations illustrate that, if predation is sufficiently strong, the predator population follows, with very fast transients, any variations in the prey carrying capacity.

4.2.2 From Donor Control to Ratio Dependence

There is a link between donor control, as illustrated in the previous section, and a ratio-dependent functional response. The aggregated, slow predator-prey model given by eqs. (4.4, 4.5) implies the following functional response:

$$g\left(\frac{N}{P}\right) = m_{12}\frac{N}{P} \qquad (4.6)$$

which is ratio dependent. Indeed, figure 1.2 shows that, as long as saturation effects do not operate (as in the present model), any ratio-dependent functional response can be approximated by

$$g\left(\frac{N}{P}\right) \approx \alpha\frac{N}{P}\left(\text{for } \frac{N}{P} \text{ small}\right) \qquad (4.7)$$

where α is the slope at the origin. If the ratio N/P is high, then the approximation used in section 4.2.1 does not apply and donor control cannot be predicted. One may expect that, in such case, predators are saturated and predation cannot be "strong" (i.e., much more important than other processes). Considering the two complementary cases of low N/P and high N/P, the full picture of a type II ratio-dependent functional response emerges (figure 1.2).

In summary, donor control has emerged from a mechanism combining spatial heterogeneity and strong predation. Strong predation is defined as occurring when predators rapidly deplete the available resource. In a homogeneous environment, strong predation leads to large oscillations that result in (stochastic) extinction of the prey. In a heterogeneous environment, however, if the prey can find refuges, the same strong predation leads to donor control. This immediately suggests where to look for donor control. If predators and their prey exhibit unstable dynamics after spatial heterogeneity has been removed (e.g., in laboratory microcosms or in field manipulations), this is evidence for strong predation. Then, according to our theoretical results, one may expect that, in the natural environments where predators and prey coexist, donor control might predominate.

Efficient predators control the abundance of their prey by eating almost all of the prey outside the refuges. However, as the available prey quickly become depleted, the predator dynamics become dependent on the slow flow of prey from the refuge. Thus, when a very efficient predator is added to a system where the prey have a refuge, the prey abundance will decline very quickly, as almost no prey that are not hidden can survive. Then predators no longer have any impact on the prey density, but their own population becomes limited by prey availability. It is clear that if a very efficient predator, which confines the prey to the refuge, disappears for some reason, a dramatic increase of the prey density would occur in the whole available space. Conversely, the arrival of a very efficient predator can result in a dramatic decrease of prey density, but, after a short transient period, the dynamics of the predator-prey system will be controlled by the donor only. This is quite a different idea from that of predators eating prey that are bound to die anyway. This model shows that there is no contradiction between donor control and the ability of predators to depress the prey density strongly.

4.3 THE ROLE OF DIRECTED MOVEMENTS IN THE FORMATION OF POPULATION SPATIAL STRUCTURES

In the simple model of section 4.2, we assumed a preexisting prey refuge and showed how its presence leads to donor control or, more generally, to ratio dependence. In the present section, we show how prey and predator dynamic spatial heterogeneity arising from predator behavior can have the same effect as a physical refuge. Locally, areas with low predator presence function as temporary prey refuges. The result is qualitatively similar to section 4.2: although Lotka-Volterra mass action can be assumed locally, the average predator density adversely affects the individual consumption when envisioned on the global scale, leading to a nonlinear predator-dependent functional response, similar to ratio dependence.

The reasons why the predator population does not distribute homogeneously in space are the taxis responses to environmental gradients. In order to explain this clustering phenomenon, we first consider in section 4.3.1 the case of a single isolated population, in which the taxis stimulus is the gradient of the population's own density (autotaxis). In sections 4.3.2 and 4.3.3, we adapt the same rationale to the situation of interacting populations: assuming that the prey density gradient is the stimulus of the predators' taxis (trophotaxis), clusters appear, with ensuing approximate ratio dependence.

4.3.1 Self-Organization Due to Accelerated Movement

The formation of spatial aggregates of animals, that is, clustering, is a common phenomenon inherent to a variety of biological species, from microscopic colonies of bacteria, insect swarms, to macroscopic fish shoals, bird flocks, and others. The fundamental mechanism of self-organization is individual taxis, which is commonly due to the spatial heterogeneity of environmental factors. We focus here on the case in which the taxis stimulus is the heterogeneity of the density of conspecifics, within a uniform external environment. Such an intrinsic mode of clustering is common, examples being swarming locusts (Edelstein-Keshet et al. 1998), midges (Okubo et al. 1977), and shoaling fish (Parrish and Turchin 1997; Parrish et al. 2002). A common observation is that clustering occurs because of the behavioral interplay between random and directional movements of individuals.

Classically, it is assumed that the taxis velocity is determined by the density gradient of some stimulus. However, it is known that diffusion-advection models that rest on this assumption cannot give a realistic representation of a cohesive group of individuals with a uniform interior density and sharp edges. Tyutyunov et al. (2004) have developed a Eulerian model of cluster formation that differs from these conventional models by making the assumption that the taxis acceleration (rather than the velocity directly) is determined by the density gradient, an assumption that is commonly used in Lagrangian descriptions of individual active behavior (Grünbaum and Okubo 1994).

This mechanistic assumption is strongly supported by the kinematic analysis of a midge swarm. In this pioneering experimental study, Okubo et al. (1977) analyzed the trajectory of swarming gall midges (*Anarete pritchardi*), measuring the trajectories, instantaneous velocities, and accelerations of individuals. This study, which is unique in its level of detail, shows that the hypothesis of taxis acceleration agrees very well with the observations (figure 4.7). Projecting the midge movements onto the horizontal plane, Okubo and coauthors showed that the mean acceleration of insects at the center of the swarm is very close to zero. The center is the place where the midge density is highest; its gradient is 0; and midges fly more or less randomly. At the edge of the swarm, accelerations toward the center are highest; the midge density is lowest; and its gradient is highest. Thus, the simplest relationship between acceleration and density gradient appears to be

$$\frac{\partial \mathbf{v}}{\partial t} = \kappa \mathbf{grad} P \qquad (4.8)$$

where $\mathbf{v} = \mathbf{v}(\mathbf{x}, t)$ is the midge velocity vector at point $\mathbf{x} = (x, y)$, $P = P(\mathbf{x}, t)$ is the population density, and κ is the taxis coefficient. In order to include repulsion, this autotaxis must change direction when the density reaches the repulsion threshold P_r. The simplest way to express this is to change the taxis coefficient as $\kappa \to \kappa(1 - P/P_r)$. The dashed curves in figure 4.7 show the best fits of this modified equation to Okubo et al.'s (1977) observations with adjustment of the two free parameters κ and P_r. The agreement is excellent, with correlations between 0.71 and 0.79 between the observations and the model predictions.

Equation (4.8) is the core of the autotaxis model of Tyutyunov et al. (2004), with the modification for repulsion and the further assumption that there is a small diffusion of velocities in order to represent the forces that equalize velocities of neighbors (Flierl et al. 1999). The equation is then coupled to the following standard diffusion-advection equation, which describes the random and directed movements of the population density:

$$\frac{\partial P}{\partial t} + \mathrm{div}(P\mathbf{v}) = \delta_p \Delta P \qquad (4.9)$$

Adding reflecting boundary conditions, Tyutyunov et al. (2004) then show that the model demonstrates the formation of spatial structures, either stationary or traveling (see figure 4.8 for some examples). This can be interpreted as dynamic self-organization, like fish shoaling or insect swarming. Analytical and numerical studies show that the link between the acceleration and the density gradient (eq. 4.8) is crucial for the appearance of spatial structures. If it is assumed that the density gradient

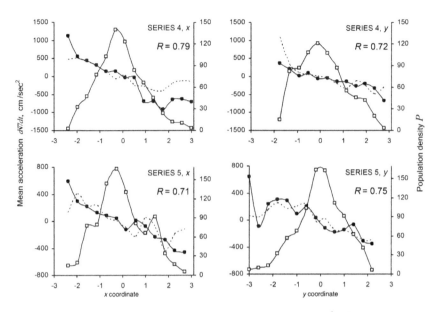

Figure 4.7. Mean values of acceleration $d\mathbf{v}/dt$ (black dots) and population density (open squares) projected onto the horizontal plane (x, y) within a swarm of midges in the Series 4 and Series 5 experiments of Okubo et al. (1977). The dashed curves show the best fit of the modified eq. (4.8) to the observed values of individual acceleration. R is the correlation between the model and the observations. After Tyutyunov et al. (2004), with permission of the University of Chicago Press. See text and original sources for further details.

determines the velocity directly, then the model does not exhibit stable heterogeneous solutions.

4.3.2 Spatially Structured Predator-Prey Systems

We now follow the same rationale as in the previous section, by adapting it to a predator-prey context (Arditi et al. 2001). The predators' taxis acceleration is now assumed to respond to the prey density gradient (trophotaxis)—not to the predators' own density. Equation (4.8) is replaced by:

$$\frac{\partial \mathbf{v}}{\partial t} = \kappa \mathbf{grad} N \qquad (4.10)$$

As in the autotaxis model of section 4.3.1, it is also assumed that velocities diffuse because of shoaling effects. This equation must then be coupled to a spatialized predator-prey model. Since our objective here is to investigate specifically the influence of active movements on the spatial distribution of predators, we assume that predator births and deaths are identically zero. In other words, we assume that, on the time scale being considered, predator reproduction and mortality are slow processes that

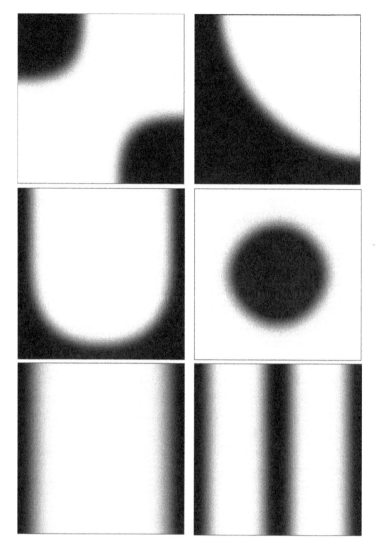

Figure 4.8. Examples of stationary spatial structures that can appear in a two-dimensional habitat for a single population obeying accelerated movement. Darker shade indicates higher population density. After Tyutyunov et al. (2004), with permission of the University of Chicago Press. See text and original source for further details.

can be neglected in front of migrations and prey consumption and reproduction. The balance equation for the predator population contains therefore only advection and diffusion terms and is identical to eq. (4.9).

Regarding the prey, we assume logistic reproduction and the familiar Lotka-Volterra functional response for predation mortality. That is, we assume that, locally, predators encounter prey according to the simple law of mass action. Accounting also for prey diffusion, the prey dynamic equation is therefore:

$$\frac{\partial N}{\partial t} = rN\left(1 - \frac{N}{K}\right) - aNP + \delta_N \Delta N \qquad (4.11)$$

The domain is assumed to be a finite rectangle with reflecting boundaries. This condition, together with the predator equation (4.9), implies that the total number of predators in the domain does not vary. In other words, the average predator density is a constant parameter P_c.

The full mathematical study of the model is given by Arditi et al. (2001). It is shown that various kinds of spatial structures can appear: uniform, periodic, quasi-periodic, or even chaotic. Figure 4.9 illustrates the spatial oscillations that can appear in some conditions. Starting from a uniform prey distribution and a patch of predators, the two populations can enter periodic spatial dynamics. The predators consume the prey locally and move simultaneously to areas of high prey density. They leave behind areas that are almost predator free, in which the prey can increase, becoming again attractive to predators, and so on. Although there are no distinct physical prey refugia, the areas with scarce predators work as reservoirs for prey production. Qualitatively, this has the same consequences as in the simple two-patch model studied in section 4.2 (figure 4.6).

In order to see this, we will now examine how the active migration behavior and the ensuing spatial heterogeneity influence the overall trophic interactions in the predator-prey system. We first define $Q(\cdot) = g(\cdot)P/N$. This function is the prey per capita predation mortality. It contains the same information as the functional response $g(\cdot)$, but, graphically, it presents very clear differences between the alternative models of Lotka-Volterra, Arditi-Akçakaya, and Arditi-Ginzburg, when plotted against P, with constant N (figure 4.10).

We now define

$$Q_\infty = \lim_{T \to \infty} \frac{1}{T} \int_0^T \frac{\overline{aNP}}{\overline{N}} dt \qquad (4.12)$$

where $\bar{x}$ designates the spatial average of x. We recall that, in our present model (eq. 4.11), the local prey consumption is simply the Lotka-Volterra expression aNP. The descriptor Q_∞ is therefore the asymptotic average in time of the spatially averaged calculation of the prey per capita mortality due to predation. It is defined in the asymptotic regime, after the prey transient dynamics have vanished. Therefore, Q_∞ only depends on the model parameters, including P_c. In the trivial case where both population distributions are stationary and uniform, we simply have.

$$Q_\infty = \frac{aNP}{N} = aP \qquad (4.13)$$

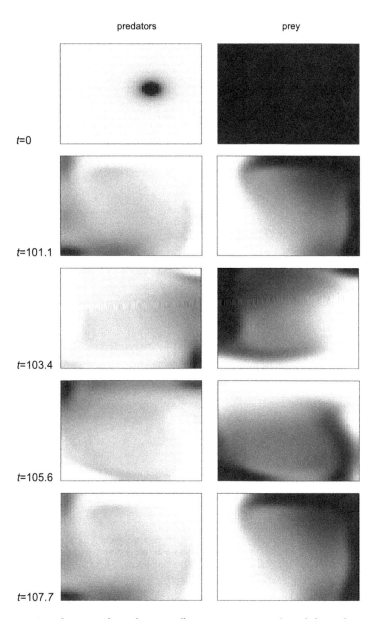

Figure 4.9. Two-dimensional population oscillations appearing in a bounded two-dimensional domain, starting from a nonsymmetrical predator distribution and a homogeneous prey distribution. Darker shades indicate higher densities of predators (left column) and prey (right column). From Arditi et al. (2001), with permission of Elsevier. See text and original source for further details.

We next study (by simulation) the way in which Q_∞ (defined by eq. 4.12) changes when the model solutions are nontrivial heterogeneous distributions of the kind shown in figure 4.9. Examples are given in figure 4.11 for several values of the taxis coefficient κ. One can see that, even though the local trophic interactions follow the Lotka-Volterra model, the averaged rate Q_∞ is

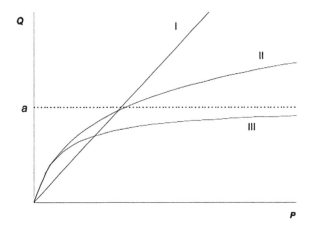

Figure 4.10. Prey mortality rate Q (due to predation) as a function of predator density P. (I) Lotka-Volterra model $Q = aN\ P/N = aP$ is linear. (II) Arditi-Akçakaya model: $Q = \alpha P^{1-m}/(1+\alpha h N P^{-m})$ ($m<1$) is strongly nonlinear, growing to infinity. (III) Arditi-Ginzburg model: $Q = \alpha/(1+\alpha h N/P)$ is increasing, tending to an upper asymptote. Representing empirical values of Q as a function of P is therefore a means of selecting among these three models. From Arditi et al. (2001), with permission of Elsevier.

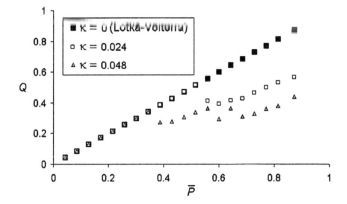

Figure 4.11. Average prey mortality rate Q_∞ as a function of the spatially averaged predator density $\bar{P}$. As the taxis coefficient κ increases, the response becomes typical of predator dependence (compare with figure 4.10). From Arditi et al. (2001), with permission of Elsevier.

completely different from eq. (4.14) when κ is sufficiently high to generate heterogeneous oscillatory dynamics. The higher the value of κ, the more pronounced is the deviation of Q_∞ from the straight line corresponding to the homogeneous steady state (i.e., the extension to the whole habitat of the Lotka-Volterra point model). From the examination of figure 4.11, it appears that, for high values of κ, the descriptor Q_∞ exhibits a saturation that, among the three cases represented in figure 4.10, is closest to the Arditi-Ginzburg ratio-dependent response. Arditi et al. (2001) also show that this result is even clearer and more precise in the one-dimensional version of the model, which is computationally much faster and more stable.

4.3.3 How Ratio Dependence Emerges From Directed Movement

The explanation for the emergence of ratio dependence is the following. A clustered spatial structure arises as a consequence of the high migrating activity of predators. This structure can be considered as an ensemble of two types of moving zones: some with high predator density and extremely low prey density, others with high prey density and a virtual absence of predators. Either passive diffusion or displacement of the zones themselves bring to the zones of the first type an amount of prey proportional to the prey population in the zones of the second type (which is almost equal to the total prey population). All such prey, entering zones of high predator density, are consumed almost immediately. Therefore, the population-level functional response, which is, by definition, the total prey consumed divided by the total predator population, will be proportional to the ratio of the total prey and total predator populations.

Thus, the model of section 4.3.2 illustrates again our suggestion that ratio dependence is a simple way of summarizing the effects induced by spatial heterogeneity, while the prey-dependent form (e.g., Lotka-Volterra) is more appropriate in homogeneous environments. This confirms the results of the previous sections in which the causes of clusters were very different. In the experimental study in which cladocerans were forced to different spatial distributions (section 4.1), we showed that homogeneously distributed populations presented characteristic features of prey-dependent functional responses while heterogeneity induced features characteristic of predator-dependent functional responses (ratio-dependent or not). In the simple model of section 4.2, the prey refuge was fixed but the results were very similar to those of the present model: prey migrating out of the refuge are consumed almost immediately and the per capita consumption rate is necessarily proportional to the ratio of prey and predator densities. In the present model, the refuge is not predetermined. The subdivision into two types of zones arises spontaneously, with zones of low predator density acting as refuges for the prey. Moreover, the structure is not spatially stationary: zones are in constant movement.

Note that trophic transfer is not necessarily required for the appearance of spatial structures in predator-prey systems. Tyutyunov et al. (2007) show that coupled movement rules can be sufficient: if the predators pursue the prey and the prey flee from the predators, population clusters can appear even in the absence of consumption, reproduction, and mortality. The condition is that both populations follow acceleration rules similar to eq. (4.10).

In a theoretical study, Cosner et al. (1999) investigated the effects of various modes of spatial grouping, including nonuniformity of the landscape itself and presence of refuges. They found that ratio dependence can arise if either the predators or the prey form clusters. Other types of functional responses (e.g., the DeAngelis-Beddington model) can also be obtained with other grouping scenarios (see also Cantrell and Cosner 1991) or other migration scenarios (Michalski et al. 1997). A model specifically developed for snail-periphyton interaction in streams also shows the effect of predator behavior leading to ratio dependence when averaged spatially (Blaine and DeAngelis 1997). The heterogeneous solutions obtained with the partial differential equation model of section 4.3.2 are similar to those demonstrated by other authors who used cellular automata or individual-based models (see, e.g., spatial chaos and shoaling behavior in Comins et al. [1992] and Flierl et al. [1999]). The problems of spatial averaging (aggregation) and of correct translation of the dynamics of a distributed population with density-dependent disturbances from one spatial scale to a finer or coarser modeling scale were investigated by Pascual and Levin (1999). In sum, whatever their nature and causes (environmentally forced or behaviorally induced), spatial and temporal heterogeneities make the aggregated functional response deviate from its local counterpart and increase the viability of the prey population and of the whole community.

Note that this occurs because heterogeneities in prey vulnerability appear. Other types of spatial heterogeneity, like the presence of obstacles (which do not offer subareas of differing quality), do not have the same consequences. The experimental study of Hauzy et al. (2010) even suggests that obstacles tend to reduce predator dependence because they reduce the encounter rate between predators.

4.4 RATIO DEPENDENCE AND BIOLOGICAL CONTROL

4.4.1 The Biological Control Paradox

The agricultural practice of controling insect pests by natural enemies presents the following paradox (Arditi and Berryman 1991). On one hand, the biological system can be quite simple, much simpler than a natural ecosystem: a single crop species growing in a uniform field, a single insect pest population specialized on this plant, and the introduced natural enemy, specialized on this pest species. Many successful cases exist, with pests maintained at densities less than 2.5% of their carrying capacities (Beddington et al. 1978). An adequate mathematical model for a pest-enemy system must exhibit persistent dynamics at such low pest density. It is not necessary for the model dynamics to be an equilibrium.

Pest extinction or fluctuations that do not exceed the economic threshold are also compatible with satisfactory control (Murdoch et al. 1985).

On the other hand, simple predator-prey models built with prey-dependent functional responses predict that it is impossible to maintain the prey population much lower than the carrying capacity (see, e.g., the Rosenzweig-MacArthur model presented in chapter 1). In such a model, the pest is represented by the prey and the enemy by the predator; the crop plant sets the prey's carrying capacity K. The paradox arises from the structure of such a predator-prey model, which gives rise to a humped (parabolic) prey isocline and a vertical rectilinear predator isocline (figure 1.1). Because of this latter property, the prey equilibrium is independent of prey parameters, as was shown in chapter 1. Its density and stability are entirely dependent on predator characteristics, with efficient predators creating low, unstable prey equilibria and large oscillations that periodically bring the prey population to values close to its carrying capacity (figure 4.12a).

The paradox is resolved if the predator isocline is slanted rather than vertical. Slanting isoclines are obtained if the functional response is ratio dependent or, more generally, if the per capita rate of increase of predators declines with their own density. With this isocline structure, the prey equilibrium can be reduced and its stability increased by changing prey parameters, for example, by reducing its intrinsic growth rate or its vulnerability to predation (figure 4.12b). More important, if prey parameters cannot be changed, the introduction of efficient predators can control the prey by producing the local pattern shown in figure 4.12c (see also figure 1.3c). All trajectories enter sector S3, where the populations are driven to extinction. However, when populations become small, environmental noise or pest immigration can easily bring the trajectories back into sector S1, where both populations grow again. Thus, the graphical properties of this model make three qualitative predictions that are compatible with the patterns that are typical of populations under biological control: (1) both populations are maintained at low densities; (2) no stable equilibrium exists; and (3) local populations experience repeated quasi-extinctions.

4.4.2 Trophotaxis and Biological Control

Sapoukhina et al. (2003) have developed a more detailed mechanistic model that explains biological control with taxis rules similar to those of the model of section 4.3. This model draws on the large body of theoretical ecology that has developed with the purpose of understanding how the natural enemy is able to keep the pest density low enough to avoid

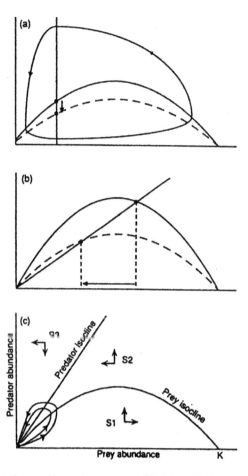

Figure 4.12. Biological control in predator-prey models. The dashed curve shows the effect of increasing prey vulnerability or lowering its rate of increase. (a) In a prey-dependent model, the effect is paradoxical: the prey equilibrium remains the same because the predator isocline is vertical; low prey equilibria are unstable and give rise to limit cycles. (b) In a ratio-dependent model, both trophic levels decline in the same proportion because the predator isocline is slanted; low stable equilibria are possible. (c) Nonequilibrium biological control in a ratio-dependent model: for very efficient predators, the isoclines intersect at the origin only; the populations remain at low densities, exhibiting growth phases followed by quasi-extinctions. After Arditi and Berryman (1991), with permission of Elsevier.

outbreaks above the acceptable threshold. The major conclusion is that nonrandom search, that is, the tendency of predators to aggregate where prey are abundant, can contribute to stability (Hassell and May 1973, 1974; Hassell 1978; Free et al. 1977; May 1978; Chesson and Murdoch 1986). Other explanations are different forms of pest refuges (Beddington et al. 1978; Hawkins et al. 1993; Lynch et al. 1998), mutual interference among predators (Hassell and Varley 1969; Beddington 1975; DeAngelis et al. 1975), and local pest extinctions (Murdoch et al. 1985; Hastings 1990; Luck 1990).

Sapoukhina et al. (2003) consider that the movement of predators is induced by heterogeneity in the prey distribution and that predators tend to cluster in regions of high prey density. This trophotaxis is assumed to be the only source of the phenomena listed above, with all environmental factors assumed to be homogeneous and constant. As in section 4.3, the mathematical setting is based on reaction-diffusion-advection partial differential equations. As in the former model, it is assumed that the directed movement of predators is not determined by their velocity itself but by the velocity variation (i.e., the acceleration), which is proportional to the prey density gradient. This differs from conventional trophotaxis models, which assume that the prey density gradient determines the advective velocity directly. Besides the Okubo et al. (1977) experiments described in section 4.3, Sapoukhina et al. (2003) provide a number of additional biological examples that support the "acceleration assumption." The trophotaxis model was further substantiated by Tyutyunov et al. (2009) and Tyutyunov, Zagrebneva, et al. (2010).

For simplicity, the model is formalized in a one-dimensional domain $[0,L]$ but the results remain qualitatively valid in two dimensions (Sapoukhina 2002). The fundamental taxis equation remains the same as in section 4.3.2, with the predator acceleration proportional to the prey density gradient. The one-dimensional version of it, with the inclusion of velocity diffusion, is

$$\frac{\partial v}{\partial t} = \kappa \frac{\partial N}{\partial x} + \delta_v \frac{\partial^2 v}{\partial x^2} \qquad (4.14)$$

The local dynamics (reproduction, pest consumption, and predator mortality) are modeled as processes acting in infinitely small and, hence, homogeneous elementary portions of the domain. Thus, according to our suggestion that prey-dependent functional responses apply to homogeneous systems, it is assumed that the local predator-prey interactions follow the Holling model. The model also differs from section 4.3.2 by including predator demography. In situations of biological control of insect pests by insect enemies, it can no longer be assumed that the predator demography is much slower than the prey's. Under these assumptions, and accounting for diffusion in both populations, the dynamics obey the following model:

$$\frac{\partial N}{\partial t} = rN\left(1-\frac{N}{K}\right) - \frac{aNP}{1+ahN} + \delta_N \frac{\partial^2 N}{\partial x^2} \qquad (4.15)$$

$$\frac{\partial P}{\partial t} = e\frac{aNP}{1+ahN} - qP - \frac{\partial(Pv)}{\partial x} + \delta_P \frac{\partial^2 P}{\partial x^2} \qquad (4.16)$$

The Holling functional response implicitly includes small-scale rapid movements of predators due to local searching behavior, while the advective term in the predator equation, in which the velocity v follows eq. (4.14), explicitly describes the directed movement of the population density on the larger and slower spatiotemporal scale. Both population life cycles are assumed to occur on similar scales, and both population densities vary in time. This is the major difference with the earlier model of section 4.3.2.

Note that the nonspatial analogue of this model is simply the well-known Rosenzweig-MacArthur model. With reflecting boundary conditions, the spatialized model also possesses spatially homogeneous dynamic regimes that correspond to those of the Rosenzweig-MacArthur model. The nontrivial uniform equilibrium corresponds to coexistence of prey and predators in the absence of predator active movements. If there are no spatial effects, it is well known that, with efficient predators (large a), a homogeneous limit cycle C_0 arises around the equilibrium.

Sapoukhina et al. (2003) have studied the influence of the taxis coefficient κ. Starting from a homogeneous stable solution, active movements make inhomogeneous spatial solutions appear. With the increase of κ, the dynamics become more complex and more distant from the original equilibrium. What is particularly interesting is that the taxis activity has the same effects on the homogeneous limit cycle C_0. Starting from this spatially homogeneous oscillating solution, the increase of κ makes it lose its stability monotonically, and new spatially inhomogeneous oscillations R_κ arise. The temporal variations of the spatially averaged densities $\bar{N}$ and $\bar{P}$ are much smaller than those of the homogeneous limit cycle C_0.

Regarding the biological control paradox, let us study a numerical example with a very voracious predator (figure 4.13). In the absence of taxis (i.e., $\kappa = 0$), the stable homogeneous limit cycle C_0 has a very large amplitude, a typical outcome of the Rosenzweig-MacArthur model. For a sufficiently high value of the taxis coefficient κ, the cycle becomes unstable and spatial patterns arise. Figure 4.13 compares the cycle C_0 and the spatially heterogeneous dynamics R_κ for $\kappa = 0.5$. The variations of the spatially averaged population densities in the regime R_κ are dramatically smaller than those of the homogeneous cycle.

Sapoukhina et al. (2003) also show that the spatial averages $\bar{N}$ and $\bar{P}$ do not hide local population outbreaks: the amplitudes of local pest fluctuations are much lower than their homogeneous counterparts at every point in space. This is a major progress over other models in which the reduction of average prey amplitude is caused by spatial desynchronization of local predator-prey oscillations, with no significant reduction of local prey outbreaks (e.g., Jansen and De Roos 2000). In other words, our model shows that the predator-prey system can be stabilized at both the global and local spatial scales. The pest density fluctuates with a very small amplitude around

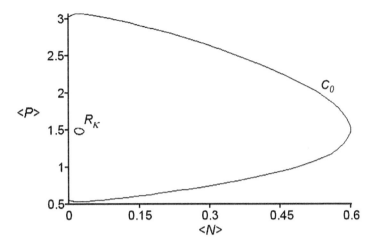

Figure 4.13. The cycle C_0 is the standard Rosenzweig-MacArthur limit cycle that appears homogenously in space in the absence of spatial effects. Taxis destabilizes C_0 and generates much smaller cycles such as R_κ in the plane of spatially averaged population densities. From Sapoukhina et al. (2003), with permission of the University of Chicago Press.

average values that are lower than 2.5% of the pest carrying capacity. High searching efficiency, short handling time, and directed movement are desirable attributes of a biocontrol agent. While high values of a and low values of h lead to low average pest density, the predator's taxis activity can stabilize the dynamics. The coefficient κ should be high enough to generate spatial patterns but not too high in order to avoid chaotic dynamics.

Qualitatively, the present model works very similarly to the model of section 4.3.2. The spatial dynamics of the interacting populations display constantly moving density patches, which agrees with field observations (e.g., Winder et al. 2001). Predators respond to the heterogeneity of the prey density by accelerating toward the spots where prey are abundant, resulting in predator aggregation. When reaching a local maximum of prey concentration, predators decelerate because the prey gradient reverses. Predator aggregations lead to local prey extinctions, while patches with low predator densities play the role of partial refuges where prey densities grow. Then predators move actively to the newly formed prey clusters. In sum, this spatially explicit model is able to capture various kinds of heterogeneity that are liable to promote the persistence of predator-prey interactions and are known to be potential explanations of effective biological control (e.g., Hassell 1978, 2000; Hassell and Anderson 1989).

Let us recall that we showed in section 4.4.1 that, among spatially independent models, the ratio-dependent approach resolved the biocontrol paradox. The results of the present spatially explicit model, which assumes a prey-dependent functional response locally, corroborate once more our conjecture (Arditi and Ginzburg 1989) that ratio dependence emerges at the global scale from any kind of local predator clustering.

4.5 EMERGENCE OF GRADUAL INTERFERENCE: AN INDIVIDUAL-BASED APPROACH

In chapter 1, we acknowledged that ratio dependence cannot hold universally. We expect it to be typical of medium to high predator densities. Actually, this was an implicit assumption of all models presented earlier in this chapter (sections 4.2–4.4). The gradual interference hypothesis that we introduced in section 1.6 suggests that, at low predator density and locally, the law of mass action can hold reasonably well, with the associated prey-dependent functional response. As the predator densities are gradually considered to be higher, mutual interference must become more frequent and lead ultimately to ratio dependence, after some transitory range. Graphically, this translates into a predator isocline that is vertical at low density and slanted at high density (figure 1.5).

By encompassing the reasonable characteristics of each model and eliminating those properties that produce biologically unrealistic predictions, the gradual interference functional response has the potential to reconcile long-standing debates. It would defuse much of the controversy associated with choosing between either the prey-dependent or the ratio-dependent functional responses.

In this section, we present two models that can lead to gradual interference, each based on a different mechanism. In the first model, the focus is on the predators' individual home ranges: as home ranges overlap increasingly, interference becomes gradually more intense. In the second model, the focus is on spatial patterns arising from animal movements, in the spirit of the model of section 4.3; the increasing spatial heterogeneity that results from movements leads to increased predator dependence in the functional response. The fact that these two rather different models of collective behavior both lead to gradual interference suggests that this must be an emerging property, arising from a variety of possible mechanisms (see section 5.5 for a further discussion of emerging properties).

4.5.1 A Qualitative Model Based on Predator Home Ranges

Ginzburg and Jensen (2008) have presented a verbal argument to capture both the prey-dependent and the ratio-dependent domains in one model. The argument draws on considerations about predator density and also about the fact that the appropriate time scale of the demographic processes depicted by differential equations is slower than the behavioral time scale on which consumption is commonly represented.

One way of understanding how differing time scales determine the functional response is to consider a simple spatial depiction of predator consumption. For the purpose of simplicity, we assume that: (1) prey are uniform in density over a two-dimensional habitat; (2) each predator individual behaviorally minimizes overlap between its own search area and the search areas of conspecifics; and (3) predators with overlaps in their home ranges show a reduction in consumption rate due to sharing available prey with their neighbors (indirect rather than direct interference). We assume that, for longer periods of time, predators search larger areas (home ranges) and orient themselves to minimize sharing of potential prey within these areas (figure 4.14). When the appropriate time scale of population dynamics is nearly instantaneous, the search area is nearly zero (figure 4.14A–C). Predators cannot interfere with each other, regardless of whether predator densities are low (figure 4.14A), medium (figure 4.14B), or high (figure 4.14C). The predator isocline that best represents the system is vertical for all reasonable predator densities (figure 1.5a).

When the appropriate time scale of population dynamics is finite but relatively short, predator home ranges are small (figure 4.14D–F). Predators do not interfere with each other at low (figure 4.14D) and medium (figure 4.14E) densities: even with a moderate home range size, there is no overlap between home ranges. Interference emerges only at higher predator densities as home ranges begin to overlap (figure 4.14F). The predator isocline that best represents the system is vertical at low predator densities but slanted at higher predator densities (figure 1.5c). Such a system can display a mix of prey- and ratio-dependent properties. The emergence of predator dependence stabilizes the system gradually.

When the appropriate time scale of population dynamics is discrete and relatively long, predator home ranges are large. Interference emerges at almost all predator densities (figure 4.14G–I), and the predator isocline that best represents the system is slanted at all but the lowest predator densities (figure 1.5b). This system displays ratio-dependent properties in most of the density range.

We argue that few if any predator-prey systems are truly instantaneous: for most systems, consumption occurs on a time scale that is shorter than reproduction, leading to some potential for interference. As a result, the question should not be whether predator dependence emerges, but instead at what predator density it emerges. Even when the time scales of reproduction and consumption are relatively close, predators will search for prey over a discrete area during a discrete time period. What we need to

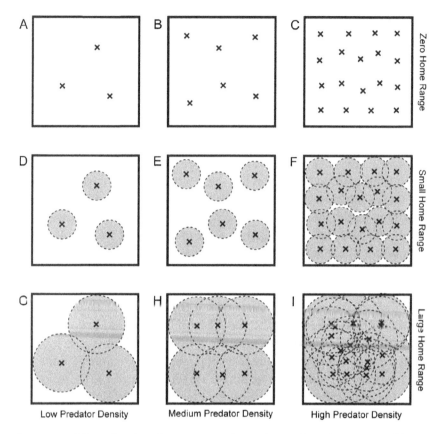

Figure 4.14. Effects of predator home range overlaps. When predators have very small home ranges, they cannot share prey regardless of whether their density is low (A), medium (B), or high (C). When home ranges are somewhat larger, prey sharing appears at high density (F), but is absent at lower densities (D, E). When home ranges are very large, some prey sharing occurs at low density (G) and can become almost complete at medium and high densities (H, I). From Ginzburg and Jensen (2008), with permission of E. Schweizerbart <www.schweizerbart.de>.

know is how densely packed predators must be to share prey. The longer the appropriate time scale of population dynamics, the larger the home range searched by predators, and the lower the density at which predator dependence emerges. For every predator-prey system, we need to assess the appropriate consumptive interval and determine how this time scale relates to predator home range.

To parameterize the indirect interference functional response empirically, the following biological data are needed: (1) the time scale of predator reproduction, (2) the area searched by predators in this time interval, and (3) the probability of capturing and consuming prey within this area. Such data are observable in many predator-prey systems.

Regardless of whether predator dependence emerges at high (figure 1.5a) or low (figure 1.5b) predator densities, the predator isocline never intersects with zero (in contrast to pure ratio dependence). This means that, in all systems, predators require a minimum density of prey to persist. This property corrects two disputable predictions of the Arditi-Ginzburg and the Arditi-Akçakaya models because, under gradual interference, (1) when predators are too far from each other to interact, prey sharing via mutual interference cannot occur; and (2) when prey density is very low, predators starve and die before encountering another prey individual (the rate of consumption falls below the metabolic rate), and predators cannot persist. This kind of minimum prey density for predator survival was first incorporated into a ratio-dependent system by Akçakaya (1992), who has offered the most parsimonious explanation of lynx-hare cycles to date (see chapter 6).

At first sight, the above rationale founded on home range overlaps does not seem to apply to territorial animals, which present homogeneous distributions. Still, territoriality is a clear mechanism of contest competition, implying resource sharing and, therefore, ratio dependence. Let us take the example of a population of birds breeding in a forest. The breeders establish territories to ensure individual resource availability. The territories cover the forest and nonbreeding floaters are pushed to low-quality habitat outside the territories. Only the territorial individuals can reproduce, but the per capita reproduction rate must be calculated with the whole population, including the shadow population of floaters. The absolute population reproduction is proportional to the total resource but, of course, the per capita reproduction rate is proportional to the ratio of the resource to the total bird population. If, in a thought experiment, we cut half of the forest and keep the same bird population, only half of the previous territories will remain available and the surplus birds will become floaters. The reproduction rate will decline to half of its previous value (with floaters counted). Of course, this only occurs if the population is high enough to occupy all possible territories. If the density of territories (hence the bird density) is so low that they are not next to each other, there is an excess of unoccupied habitat, no floaters, and we can expect to observe prey dependence. Thus, we predict again that the functional response must change gradually from ratio dependence at high predator density to prey dependence at low predator density.

4.5.2 An Individual-Based Model Based on Trophotaxis

After the PDE-based models presented in sections 4.3 and 4.4, Tyutyunov et al. (2008) have followed an individual-based approach in order to study more accurately how different assumptions about individual movements lead to the emergence of various kinds of prey and predator dependence in the population-level functional response.

The model describes explicitly the spatial movements of individuals of both populations, the encounters of predators and prey, and their demography. Keeping track of each prey and predator individual, it is possible to calculate the number of prey consumed by each predator and, hence, to evaluate the average value of the predation rate for any combination of prey and predator abundances (i.e., the functional response, by definition). Being able to separate demographic processes from spatial behavioral processes, we will show how different assumptions about the individual movements of predators and prey lead to the emergence of various kinds of functional responses at the population level.

The full details of the model can be found in Tyutyunov et al. (2008). The major assumptions are as follows. Space is continuous and time is discrete ($t = 0, 1, 2, \ldots$). The two populations consist of N prey and P predator individuals, dwelling in a closed rectangle $L_x \times L_y$ with reflective boundaries (assuming that absorbing boundaries does not change the results, qualitatively). At each time t, each individual prey i and predator j is characterized by its age, its spatial coordinates, and its vectorial velocity.

We suppose that each prey individual produces a normally distributed odor. The distribution of the odor of several individuals is simply obtained by superimposing all individual odors. Both prey and predator individuals move randomly. This is modeled by adding a random change to each individual's position at each time step. Additionally, predators are capable of directional movement, being attracted by prey odor. With the same rationale as in sections 4.3 and 4.4, it is assumed that the change in the individual taxis velocity $\mathbf{v}_j$ is determined by the prey's odor gradient and characterized by a trophotaxis coefficient κ and an inertial coefficient ν.

Any prey situated within the predator detection radius R will be eaten with probability $1 - \varepsilon$ per time step, where ε is the escape probability. Although the model could include reproduction and mortality, demographic variations are ignored in order to focus the investigations on the question of the functional response. The prey population decline due to consumption is also eliminated as follows: when a prey individual is caught, it is immediately replaced by another individual placed at a random point.

For each predator individual j, a record is kept of the total number of prey consumed since its birth, $G_{j,t}$. The individual lifelong average rate of consumption is then $g_{j,t} = G_{j,t} / \tau_{j,t}$ where $\tau_{j,t}$ is the individual's age. The population-level functional response is the average over all individual predators, that is,

$$g(N,P) = \lim_{t \to \infty} \frac{1}{P} \sum_{j=1}^{P} g_{j,t} \qquad (4.17)$$

This is reasonable and consistent because the population abundances N, P are kept constant in time. The function $g(N, P)$, so defined, was obtained numerically with Monte Carlo simulations performed for various values of N and P (see Tyutyunov et al. [2008] for details).

In the simplest case, in which neither prey nor predators are capable of directional movement, predators do not aggregate and encounter prey at random. The empirical functional response (eq. 4.17) is best fitted by the Lotka-Volterra model $g(N) = aN$. This function, which does not depend on the predator abundance P, explains 99.6% of the variance.

A more interesting situation is the case in which predators hunt the prey actively. Figure 4.15 presents the isopleth diagrams of the Monte Carlo–derived functions $g(N, P)$ for two different cases differing in the predator taxis parameters: the right panel corresponds to higher values of both trophotaxis and inertia.

The isopleths show, first, that the empirical functional responses are predator dependent; prey-dependent responses would present vertical isopleths. Second, they show that predator dependence is gradually more intense as predator density increases. This is particularly clear in the left panel: the isopleths are almost vertical at low prey and low predator densities but become progressively more slanted at higher densities. For high densities and high taxis (right panel), they are almost rectilinear, typical of ratio dependence. These features are in agreement with the gradual interference hypothesis, introduced qualitatively in section 4.5.1.

Tyutyunov et al. (2008) suggested an expression for a theoretical functional response that displays prey dependence at low density and ratio dependence at high density. This model for gradual interference was already introduced in chapter 1 (eq. 1.22). The isopleths of this model are

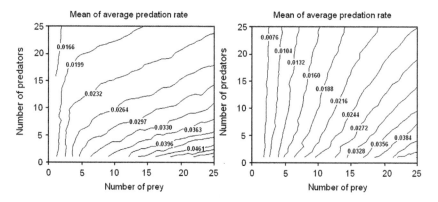

Figure 4.15. Isopleth diagrams of the empirical functional response $g(N,P)$ obtained in simulations with two sets of taxis coefficients; left: $\kappa=0.2, \nu=0.5$; right: $\kappa=0.5, \nu=1.0$. Higher taxis (at right) leads to more rectilinear isopleths. From Tyutyunov et al. (2008), with permission of Elsevier.

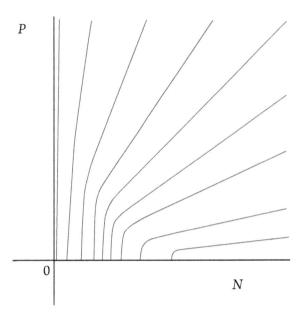

Figure 4.16. Typical isopleths $g(N, P) =$ const. for the gradual interference model (eq. 1.22). From Tyutyunov et al. (2008), with permission of Elsevier.

displayed in figure 4.16. For the two cases illustrated in figure 4.15, this model explains respectively 62.6% and 98.6% of the variance.

Tyutyunov et al. (2008) further study the more complex situation where, in addition to the predator trophotaxis described above, the prey also are capable of directional movement, sensing the predators' odor, and fleeing in response to the odor's gradient. This generates complex outcomes that we do not report here.

In conclusion, the present model, although different from the model of Ginzburg and Jensen presented in section 4.5.1, reaches similar conclusions regarding the emergence of gradual interference. Therefore, prey dependence and ratio dependence should not be regarded as competing universal models but as the extreme modes of a common model. The important biological question is that of determining whether natural systems are close to one or the other extreme modes, or close to various intermediate modes. We believe that the experimental and natural evidence that we provided in chapters 2, and 3 suggests that, in most cases, biological systems are reasonably well described by the ratio-dependent mode of interaction. The departures from this mode lead to interesting problems, and raise interesting questions, that must be answered by fieldwork or theoretical work. Still, the "null model" for predator-prey interactions appears to be the ratio-dependent mode. This is further discussed in chapters 5, and 6.

CHAPTER 5

The Ratio Dependence Controversy

Objections to ratio dependence arose immediately following our 1989 article. The points and counterpoints of the debate have been reviewed three times (Akçakaya et al. 1995; Abrams and Ginzburg 2000; Jensen and Ginzburg 2005). In this chapter, we comment on aspects that were not sufficiently covered in the previous chapters. The controversy is substantially settled by now: a majority of ecologists accept the need to include predator dependence (i.e., interference, as we also call it) in functional response models, and ratio-dependent predation is now presented in major textbooks (e.g., Krebs 2009, 194; Gotelli 2008, 148; Begon et al. 2006, 304–305). This is a substantial change from the insistence on prey dependence as the only correct view, just 20 years ago. Our 1989 article, in which we introduced the concepts of ratio dependence, predator dependence and prey dependence, is highly cited, with growing annual citation rate.

Nevertheless, while agreeing, when pressed, on the need to include interference, many ecologists still have the reflex of using the Lotka-Volterra model (or similar prey-dependent models) when building complex food web models, either for theoretical purposes or for applied studies. Theoretical conclusions based on the prey-dependent view are still widely accepted as if they were experimental facts, with the paradox of enrichment and the cascading enrichment response of trophic chains remaining the most noticeable tenets.

The safe and certainly correct view that predator interference has a central role in ecological dynamics won acceptance largely because of our suggestion for the more radical ratio-dependent version. We are quite satisfied in contributing to move the majority's opinion in this direction. Still, we remain of the opinion that the general case is often quite close to our "null model" proposal of ratio dependence and nearly always distinctly away from the prey-dependent alternative. The empirical evidence presented in chapters 2 and 3 brings much support to this view.

To keep the present book focused on predator-prey theory, we have chosen not to review models of host-pathogen dynamics. Still, it is interesting to stress that a similar dichotomy exists in dynamic epidemiological models, where the analogue of prey dependence is called *density dependence* and the analogue of ratio dependence is called *frequency dependence* (e.g., Anderson and May 1991; Cook and Busenberg 1993). Among epidemiologists, this question is not a matter of controversy. Density-dependent transmission is generally considered to occur in pathogens dispersed randomly by air in great numbers over great distances, while frequency-dependent transmission appears more reasonable where each infected individual comes into contact with an approximately fixed number of susceptible hosts, as in many human diseases with person-to-person transmission (see Holt [2011, 151–156] for a nice short review).

5.1 EVIDENCE OF RATIO DEPENDENCE IS OFTEN CONCEALED IN THE LITERATURE

The review of data analyses that we made in chapter 2 has shown that direct estimates of functional response parameters and direct identification of functional response models are not trivial quantitative procedures. Because of the fundamental nonlinearity of saturating functional responses, the mutual interference coefficient m can only be estimated in a reliable way if trivariate measurements are available: the dependent variable (the number of prey eaten per unit time) must be estimated for various values of prey density and various values of predator density (the two independent variables). As explained in section 2.1, the application of the original method of Hassell and Varley (which ignores saturation) leads to underestimated values of m. Another source of difficulty, which led to errors in some publications, is the fact that fitting the Arditi-Akçakaya model requires solving numerically an implicit equation (eq. 2.6).

When corrected for the presence of saturation, the m values of table 2.1 increased from a typical 0.4 to a typical 0.75. Very crudely, one may extrapolate that in large data sets, m values calculated with the Hassell-Varley method should be corrected by a factor of almost 2. Of course, this cannot be a corrective factor applicable to every single study, but it could be used as a fast and dirty correction in those cases lacking the data required by the Arditi-Akçakaya method such as, for example, the data set of Hassell (1978, 91).

A number of articles contain evidence of ratio dependence (i.e., $m \approx 1$) but do not make this message clear. An example is the meta-analysis of Skalski and Gilliam (2001). As we showed at the end of section 2.1.4, the strong support that this article brings to the Arditi-Akçakaya and to the Arditi-Ginzburg models appears only when examining the online appendix.

Finally, a lesson given by the reanalyses of section 2.1 is that the correct identification of ratio dependence or the quantification of mutual interference are very sensitive to subtle mathematical considerations. Quite different conclusions are drawn when altering some assumptions or waiving some approximations that, on first sight, could appear acceptable.

5.2 THE PARADOX OF ENRICHMENT AND THE CASCADING ENRICHMENT RESPONSE: IS THERE ANY EVIDENCE THAT THEY EXIST?

Evolutionary elimination of unstable configurations is a potent force that has shaped many aspects of the natural world. The solar system as we see it today is stable because all unstable three-body configurations were eliminated: various bodies fell into the sun or into each other, forming the stable nested structure of planets and their satellites (Arnold 1990). Likewise, ecological systems have been subject to such instability elimination, which has been suggested multiple times (e.g., May 1976; Colyvan and Ginzburg 2010; Ginzburg et al. 2010).

To see the picture clearly, let us start with a single species model. An equilibrium in a density-independent model is possible if the birth rate is exactly compensated by the death rate. Such equilibrium is, however, structurally unstable. That is, a small mutation in one of the rates will either (1) lead to extinction or (2) lead to an exponential growth, which will be stopped by some limiting factor (food, space, light, etc.). So the only equilibria we expect to see are the ones driven by density dependence underlined by a set of specific limiting factors. Andrewartha and Birch (1954) held a strong density-independent view (random walks of abundance, away from resource limitations). This view has generally lost and today we think that only structurally stable equilibria are to be seen (Begon et al. 2006).

The situation is only slightly more complex for the predator-prey pair system. At the prey-dependent extreme, the equilibrium is quite possible; it can even be stable with respect to initial conditions (Lyapunov stability), but it is structurally unstable. The root of this lies in the paradox of enrichment (Rosenzweig 1971). Improvement in predator efficiency or improvement in prey carrying capacity or reproductive rate lead to instabilities followed by extinction. So, if isoclines cross as on figure 1.5a, that is, in the prey-dependent part of the predator isocline, extinction is highly likely. If they, however, cross as on figure 1.5b, away from the prey-dependent limit and closer to the ratio-dependent limit, the equilibrium is structurally stable. Equilibrium in this case changes little if parameters of either interacting species evolve in any direction.

Thus, what happens with the interacting pair is analogous to the situation with a single species. We are not likely to see extant pairs of predators and prey close to the prey-dependent limit, even though it is logically possible, and even though such pairs might be assembled in laboratory experiments. Structural instability will tend to eliminate these cases, and only those with some degree of interference will survive.

Rosenzweig clearly understood the issue of elimination of unstable configurations and wrote a little-known article where he raised this possibility as a problem (Rosenzweig 1973, 85). He then developed a coevolutionary argument according to which the system "may not become extinct" (p. 93). This argument is not very convincing, and it is our opinion that the empirical evidence for Rosenzweig's paradox of enrichment is lacking (Jensen and Ginzburg 2005; and see below). The reason may simply be that unstable configurations of a prey-dependent type (as in figure 1.5a) are selected away. The structural stability of ratio-dependent equilibria (as in figure 1.5b) may be the reason why most practical evidence points to these kinds of equilibria. In the three previous chapters we have shown that this is, indeed, where we find sustainable predator-prey systems in nature. To Rosenzweig (1971), the effect looked paradoxical because an unpleasant outcome (instability, leading possibly to extinction) stemmed from resource enrichment, which on first intuition should be beneficial to the ecosystem and not cause any harm. The reason this sounds paradoxical to us is that it is very likely that this effect is not a true phenomenon but the (correct) mathematical result of an inappropriate model. The nonexistence of both paradoxes is discussed in detail in chapter 3.

By nonexistence, we mean here nonexistence in nature. In the laboratory, a number of skilled biologists have applied remarkable efforts to devise experiments intended to "confirm" existing theories. Such efforts are often akin to attempts to build "biological computers" on the model of the analog computers of the 1940s–1960s, which mimicked differential equations with electric circuits of resistors, capacitors, and inductors. Gause was the first to yield to such temptation (Kingsland 1985, 150). As we mentioned earlier, in his first experiments with freely reproducing populations of *Didinium* (predator) and *Paramecium* (prey), he could not prevent the latter from incurring extinction (Gause 1934a,

1934b). So he resorted to reigniting the cycles by adding prey from the outside. That the cycle period did not depend on the frequency of the additions was then interpreted as a confirmation of the prevailing Lotka-Volterra theory. In later experiments performed with yeast as prey and *Paramecium* as predator, he obtained persisting cycles by controlling the yeast growth rate and the *Paramecium* dilution rate (Gause 1935a, 1935b). Of course, ratio-dependent theory would be even better confirmed by these experiments. It would explain both the extinctions and the period independence (see section 5.4).

Likewise, more current literature reports quite ingenious experiments that, with some arguments, can be interpreted as confirmations of the paradox of enrichment (Luckinbill 1973, 1974; Veilleux 1979; Holyoak 2000) or of the cascading response (Kaunzinger and Morin 1998). In the laboratory, one is certainly capable of creating prey-dependent interactions with sufficiently rarefied consumers and then producing what is

Georgii Frantsevich Gause (b. Moscow 1910; d. Moscow 1986). As a young student, this Russian biologist developed a great interest in American science and particularly in the demographer Raymond Pearl. He argued that fieldwork, with too many variables, could never adequately explain ecological relationships, and that only in the simplified laboratory environment, where variables could be controled, would it be possible to determine precisely how a specific ecological factor influences a population. He was only 24 when he published *The Struggle for Existence* (1934b) in the United States, which presented his now-famous competition and predation experiments with protozoa. The next year, he published in France a second, less-known book that included additional predation experiments, with the explicit aim of testing the Lotka-Volterra theory: *Vérifications expérimentales de la théorie mathématique de la lutte pour la vie* (1935). As a professor at Moscow State University, his later scientific life was devoted to the development of antibiotics. See Kingsland (1985) and Wikipedia for additional historical information. Drawing by Amy Dunham.

interpreted as confirmation of the standard theory. However clever, these experiments suffer the same epistemological problem as those of Gause or, in the words of Hutchinson (1978, 23): "What we have indeed done is to construct a rather inaccurate analog computer for giving numerical solutions of our equation, using organisms for its moving parts." Are these experiments representative of what occurs in nature? All the natural data that we know point in another direction.

As we have shown in chapter 3, the trophic cascade of equilibrial responses to suppression of the top predator is not only a well-established fact but a logical consequence of both prey-dependent and ratio-dependent theories. In this respect, the situation is analogous to the pulse responses: cascading irrespectively of the adopted model. A striking difference exists for the responses to enrichment at the bottom of the chain: it is monotonic for a ratio-dependent chain and alternating, top-down cascading (actually, a bit more complex) in a prey-dependent chain. We have shown in chapter 3 that the monotonic response is a much more reasonable model of what is seen in data.

5.3 THE FALLACY OF INSTANTISM

The root of the disagreement between advocates of prey and ratio dependence, as is often the case, lies in a seemingly orthogonal, and somewhat philosophical, direction. The issue concerns the interpretation of the rate dN/dt in population growth equations. Most people in the prey-dependence camp (including Lotka and Volterra themselves) take the growth rate represented by the derivative dN/dt in the relevant differential equations to be truly instantaneous. We call this view *instantism*, a term coined in Ginzburg and Colyvan (2004, 70–74). We already pointed out the essence of the argument in Arditi and Ginzburg (1989) with a telltale quotation of Rosenzweig and MacArthur (1963). In recognition of the fact that predator density should adversely affect the number of predators being produced, these authors stated: "the greater the number of predators, the faster the [prey] density is reduced, still the instantaneous rate of change of the predator population depends only on the instantaneous rate of kill, which depends on the instantaneous density of prey." Restating the same idea, Rosenzweig (1977) explained that the adverse effect of predator density on predator growth transits through prey dynamics. This argument holds only if all "instantaneous" values refer to truly continuous variables varying on the same time scale—if an instant of feeding is of the same order of magnitude as an instant of reproduction or mortality.

When predators reproduce once a year but consume prey every day, the adopted abstraction is to treat the annual reproduction as the result of the corresponding daily rate, thought to be instantaneous. Turchin and Batzli (2001), in their model for arvicoline rodents, presented a set of parameters in annual units. When one of us (LRG) wondered what the simulation time step was, the answer was that it was one day: the daily reproduction rate of these mammals was set by dividing the annual value of r (say, 4–6 offspring per year) by 365. This rate was then used in simulations. This approach is so ingrained in the instantist view that these authors did not even explain how it was done: all rates are "instantaneous." With this abstraction in place, prey-dependent predation is natural because, in an instant, a single consuming predator cannot react to whether there are other predators nearby—the predator in question only responds to the local prey density.

A common argument for instantism is that reproduction occurs at every moment—it does not occur in discrete generational time steps. For instance, in a laboratory bacterial culture, at any given instant, there are some cells reproducing, even though the generation time may be about an hour. And in the case of a human population, at any given second, there are humans reproducing. This simple observation suggests that a shorter or even an instantaneous time scale is the most appropriate. In the case of exponential growth, this argument seems sound because the age (or stage) structure equilibrates and an instantaneous view is a possibility. But, as we have stressed repeatedly (Ginzburg and Colyvan 2004), the proper subject of ecology is deviation from exponential growth, and from this vantage point things are rather different. For example, when the food supply changes, the response time of bacteria, with 20-minute generation times, is very different from that of humans, with 20-year generation times. The common argument for instantism is thus without force. In our view, the dt in dN/dt should be thought of as a large, finite interval that includes both the reproduction and consumption events. Only at such a scale can we sensibly include, for instance, the conversion of food into offspring (Ginzburg 1998; chapter 1 of this book). In our view, the scale for defining various rates is different in different cases, but it is never truly instantaneous—to assume so is to read too much into the mathematical formalism. Getz (1998, 547–548), in an excellent review of the art of ecological modeling, covers the instantist misconception in ecology quite well.

Even in physics, where differential equations are of fundamental importance, we know that the underlying assumption of continuity is an idealization. Liquids, for example, are not continuous; they consist of distinct molecules. Even space-time may not be continuous, but may come in very small, discrete packets. And in cosmology, Einstein's equation treats the distribution of matter in the universe as continuous, when clearly it is not: stars are quite distant from each other. For many physical applications, derivatives should thus not be thought of as literally instantaneous rates, but rather as idealizations of finite rates understood at the appropriate scale.

In ecology, this point is even more important than in physics. We must acknowledge that the ecological mechanisms are very complex and cannot be described in every detail, instant by instant. We believe that, if the emerging phenomenon is observed at an appropriate finite time scale, the effect of predator interference will become more noticeable because it is more pronounced at longer time scales.

In section 2.4, we showed that the time series of data for wolves and moose on Isle Royale revealed a ratio-dependent interaction. On the whole island scale, the ratio-dependent Arditi-Ginzburg (AG) model was able to account for 64% of the kill rate when viewed annually. This was substantially higher than the 35% of the prey-dependent Holling (Ho) model, which is of equal complexity. By taking moving averages of all variables over two, three, or more years, one mimics different observation time scales. Figure 5.1 shows the results of fitting the AG and Ho models to such data. The prey-dependent Ho model is outperformed by the AG ratio-dependent model. With the generation time of wolves close to 4 years and that of moose around 9 years, it is interesting to note that fits reach maximal values with a window size of 5–6 years, where the AG model accounts for about 90% of the kill. Although the prey-dependent model also improves with temporal averaging, the explained percentage of the kill rate remains much below that of the ratio-dependent model.

Fussmann et al. (2005) observed prey dependence in a laboratory experiment with rotifers. Jensen et al. (2007) reanalyzed it and reinterpreted it along the line of instantism. The time scale of the experiment was too short to observe any consumer interference: trials lasted 4 minutes, less than the gut passage time. In a way, this is another example of a clever experimental work that mirrors the prevailing theory in a fashion similar to a biological computer, as discussed in section 5.2. Interestingly, other experiments with rotifers performed with longer interaction times (30 minutes) support ratio dependence (Tyutyunov, Titova, et al. [2010], an article unfortunately available in Russian only).

Our view, if taken literally, calls for difference equations, where time is treated discretely. These equations, however, are notoriously hard to deal with mathematically. We therefore continue to use differential equations, but we bear in mind that these are idealizations of the underlying finite, discrete-time processes. Building simple crude models of complex processes justifies this abstraction, which has been common in other fields of

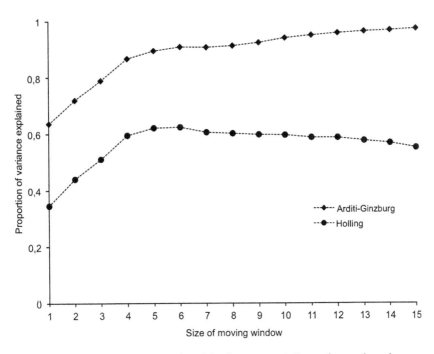

Figure 5.1. Proportion of variance explained for the per capita kill rate of moose by wolves as a function of the size of the moving average window. Averaging over 5–6 years appears to be reasonable. After Jost et al. (2005), with permission of Wiley-Blackwell.

sciences and used with the same justifications. Instantists, on the other hand, treat differential equations literally, as the representation of truly continuous processes. The conflict between these two interpretations of differential equations is a conflict between two abstractions. Both have their problems. We have already made clear the problems with instantism, but our discrete-time interpretation is not free of problems either. After all, there are infinitely many different-sized, finite time intervals to choose from, and none, it would seem, is the privileged "correct" one. When we choose the "appropriate" one, as we have suggested, we too are making an idealization. Which of these two idealizations is preferable? Time will tell.

Instantism, as described above, is a special case of a larger problem that can be termed *literalism* in applying mathematical constructs to biology (Jensen and Ginzburg 2005). A good example is the partitioning of actual biological species into discrete and continuous categories according to their demography, an approach followed by almost all textbooks. Discrete species are said to be those that reproduce every spring, for instance. Other species, like humans or bacteria, are said to be continuous: they reproduce at any time without a specified frequency. Differential equations are then said to be the correct abstraction for continuous species and difference equations the correct abstraction for discrete species. We find such a literal projection of the mathematical construct onto biology very restrictive and unreasonable. Depending on the time scale appropriate to the particular case, the same species can be described by both constructs. Darwinian natural selection is a prime example. It does not matter whether a geometric series (discrete) or exponential growth (continuous) is invoked to deduce the idea of natural selection from the Malthus law. The law is just a caricature that captures the necessary properties of the process in either form.

To use an analogy from art, we feel that literalist theorists are attempting something akin to photorealism. We view good theoretical work as akin to impressionism. Rather than interpreting our mathematical constructions as literal depictions, we see them as metaphors. For instance, we use differential equations but allow for mechanisms such as particular forms of predator dependence that emerge in time steps that are larger than instantaneous. Certainly, inclusion of such mechanisms violates literalist rules, but the resulting metaphor works better, as we have demonstrated in chapters 2, 3, and 4. In relaxing literalist constraints, we hope to make theoretical work more practical and tractable—the goal is to produce rough but meaningful models (Ginzburg and Jensen 2004). We see the metaphorical approach to modeling as being far more realistic than the literalist alternative.

5.4 HOW THE RATIO-DEPENDENT MODEL SERVES THE DEBATE ON THE CAUSES OF CYCLICITY

As is common in all areas of science (and, really, beyond science) the majority's opinion on unresolved matters is much more unanimous than that of the specialists. If you ask typical ecologists about the cause of population cycles, predator-prey interactions will come up as a dominant response. This is because it is taught in every textbook. The simultaneous publication in the 1920s of the Lotka-Volterra model and the description of vertebrate cycles by Elton planted the association in the ecologists' imagination and in ecology's canon.

Predator-prey interactions clearly can account for cycles, but they are not the only explanation. If you ask a narrower group of ecologists specialized on the study of cycles, you find an unsettled debate: while about half of them favor the predator-prey explanation, the other half think that the causes of cycles lie in the intraspecific dynamics, that they are physiologically or ecologically internal to the cyclic populations. In this view, predators simply ride the wave generated internally by the prey populations. We are not about to resolve this long-standing issue here. Our purpose is to show that, no matter which side you are on, the ratio-dependent theory presents a better structure to describe cyclicity.

First, let us consider the predation hypothesis. Standard prey-dependent theory is prone to cycles. They arise over such a large swath of parameter space that one is led to consider stable equilibria an exception to the rule. In reality, cyclic species are quite rare in nature. Ratio-dependent theory accommodates this observation quite well. This is because pure ratio dependence (without the gradual interference effect) typically leads either to stability (describing a vast majority of species in nature) or to dual extinction. The latter is the case when a specialist predator consumes all the prey and dies following prey extinction. Dual extinction is avoided in the more realistic gradual interference formulation because the interaction becomes prey-dependent at low predator density. This renders the extinction equilibrium (0,0) unstable as in the standard theory. The prey population escapes low abundances and grows, followed by the predator population. The situation leads to a stable limit cycle. A major difference with the Lotka-Volterra theory is that the cycle here is a single limit cycle, not a nested set of cyclic trajectories. The cycle is also moderate in size, and it does not lead to extinction under evolutionary changes as in the Rosenzweig-MacArthur model. The lynx-hare model by Akçakaya (1992) is the best reference where the stable limit cycle is explained by the ratio-dependent mechanism supplemented by hare refuges. In this model, the effect of refuges is mathematically equivalent to the gradual interference hypothesis: dual extinction would be the outcome if no correction at low abundance was made. Gause's (1934a, 1934b) laboratory experiments are a good confirmation of the above theory. To prevent dual extinction, Gause added a sediment serving as a refuge to *Paramecia*. Alternately, he reignited the cycles by periodic prey additions. The latter effect was practically the same as refuges: dual extinction was replaced by a limit cycle.

Now, let us consider the alternative view: internal causation. There are a number of arguments in favor of this, particularly as it relates to maternal effects (Inchausti and Ginzburg 2009). Two of them are as follows:

1. The observed periods of cycles expressed in units of generation time are either below two or above six; no cycles are known with periods between two and six generation times. This turns out to be a simple prediction of a maternal effect model (Ginzburg and Colyvan 2004, 52–57), which fits observed periods quite well.
2. In cycling predator-prey pairs, cycle periods are correlated with the prey body size with a ¼ power law (the Calder allometry). Since generation times scale to body size with the same ¼ exponent, this confirms suggestion (1) above. A typical period of this allometry corresponds to

eight generations. Calder discovered the relationship in 1983, and Krukonis and Schaffer (1991) confirmed his finding using a much larger data set. A second important finding of the later article is that no correlation exists between cycle period and predator body size. These two findings, viewed together, attribute the cause of oscillations to mechanisms that are internal to the prey population.

Additional arguments in favor of the maternal effect as a cause of cycles can be found in chapters 4 and 5 of Ginzburg and Colyvan (2004).

Of course, with the internal causality view, the predator dynamics simply follow the food supply without affecting the prey cycles. This situation is quite likely if the predator-prey interactions are approximately donor controled. As discussed in chapters 1 and 3, donor control is ratio dependent. A similar dynamic does not emerge naturally from prey dependence.

The external versus internal causality of cycles is still a matter of debate. Mathematically, cyclicity demands two dimensions. These dimensions may be species abundances, or they may be the abundance and individual quality of a single species, as in the maternal effect model. Whether one sides with the predator-prey view or with the single-species, internal cause view, the ratio-dependent theory serves both sides of the debate well.

5.5 MECHANISTIC VERSUS PHENOMENOLOGICAL THEORIES

Emergent properties are simple rules acting at a collective level; they are not reducible to the mechanisms of interaction of the elements of the collective. Robert Laughlin (2005) defines emergent properties as the ones that are stable with respect to modifications of the underlying microscopic mechanisms. This general line of thought has a long tradition in physics with occasional forays into biology, sociology, and economics (Anderson 1972; Prigogine 1997). Laughlin defends the more radical view that all laws of physics have this collective origin, thus announcing the end of reductionism altogether. Whether or not collective properties are the only candidates for laws is a debatable question, but we believe that the one under consideration in this book is such an emergent property. That is, there is a variety of mechanisms that lead to it, and we have presented a number of them in chapter 4.

The term *mechanistic* comes from the attempts to explain electricity and other physical phenomena by building a literally mechanical model of them. Arguments were not accepted as true explanations unless an underlying mechanical model was built. Physics, of course, is long over this misconception. Philosophically, the distinction between a description and an explanation is not very clear. New suggestions in science are often accused of "not having a clear mechanism," this word being generally used as an underlying explanation through more elementary processes than the ones being described. In time, when people are used to a theory working well, the accusation subsides and what was viewed as phenomenological becomes mechanistic itself. Grand examples are Newton's gravitation (the previous vortex-based theories for planets, including Descartes's, were mechanistic but wrong), Wegener's tectonic plate movements, and, of course, Darwin's natural selection. Greene (2001), in a very nice short note in *Nature* titled "A Tool, Not a Tyrant," gives a set of examples from the history of science pointing to this common evolution of opinion. So we are certainly not the first to bring this issue up for discussion. Percy Bridgman wrote this in 1927: "I believe many will discover in themselves a longing for mechanical explanation which has all the tenacity of original sin. [...] Just as the old monks struggled to subdue the flesh, so must the physicist struggle to subdue this sometimes nearly irresistible, but perfectly unjustifiable desire" (quoted from Greene 2001). In ecology, Pimm (1991, 17) famously said: "one scientist's mechanism is another's phenomenon."

A number of opponents to ratio dependence asserted repeatedly that this theory lacked a mechanistic basis (e.g., Abrams 1994; Murdoch 1994). We are certainly not against finding an explanation based on elementary processes, underlying the one we are describing. The fact is that there are many possible scenarios that lead to approximate ratio dependence. In chapter 4, we have presented a number of them. Ratio dependence "emerges," in the sense introduced above, because the result is roughly independent of the detail of the underlying interactions (see the discussion in section 4.5).

5.6 "THE TRUTH IS ALWAYS IN THE MIDDLE": HOW MUCH TRUTH IS IN THIS STATEMENT?

The folk wisdom expressed in the title above is part of so-called common sense. Common sense is literally "common" and thus wrong a lot of the time. In many instances, progress in science was obtained when violations of common sense were given serious consideration. Today's TV journalists giving equal time to two sides of an issue are often doing the job of obscuring the truth. Giving equal time to evolutionary theory and to intelligent design is just one of many examples when, simply, one view is correct and there is no need for a compromise. There is no universality in the common wisdom; it works sometimes, and it fails many times.

A bit of personal history may be of interest here. In 1999, Lev Ginzburg invited Peter Abrams to Stony Brook University, where they had a public debate on the issue of ratio dependence, with Charlie Janson as the moderator. During this visit, Peter agreed to coauthor a joint article on the problem, which proved to be well read and useful. In this review, we agreed on most things, relegating our disagreements to a short final section. Peter was, and still is, the major defender of prey dependence as the null model, that is, the better starting point for modeling consumer-resource interaction. Lev's position was, and still is, that ratio dependence is a better null model. This is where the two authors disagreed. However, the disagreement is, today, much less heated than it was in the 1990s. The controversy has subsided substantially due to the joint review by Abrams and Ginzburg (2000). The majority now agrees with the "middle opinion" that predator dependence has to be included in the theory of predator-prey interactions. The disagreement is only that the authors of this book see the most frequent case to be quite close to the ratio-dependent end of the spectrum and that our opponents would rather not choose any range. So, "common sense" has helped the common opinion to change in a correct direction, but still the change is insufficient in our view.

CHAPTER 6
It Must Be Beautiful

We have borrowed the title for this chapter from a collection edited by Farmelo (2002). The collection includes chapters by theoreticians describing various basic equations in science, mostly from physics, but one from ecology (May 2002, 86–89) and one from evolution (Maynard Smith 2002). The essence of this collection is the demonstration of how mathematical beauty has served systematically as a criterion in theory choice.

The standard view is that scientific theories are deduced from the evidence. As already explained by Ginzburg and Colyvan (2004, 120), the truth turns out to be much more subtle than this simplistic picture. Of course, theories are tested by comparison to the evidence, which remains the final judge of their validity, typically under a restricted set of circumstances. The evidence, however, is often ambiguous and does not fully support or fully fail to support a particular hypothesis. Think of the standards of evidence in legal proceedings. The best and clearest evidence supports the hypothesis "beyond reasonable doubt", whereas weaker evidence might be merely "clear and convincing". Still weaker evidence is when the hypothesis is supported merely on the "preponderance of evidence". If we admit various degrees of proof, as it seems we must, then science is not simply a matter of looking at the data and producing a theory that fits the data—there are degrees of fit. Moreover, even this order of events can be questioned. Very often, we have a theory first, and the theory suggests what data we ought to seek and how we should interpret it. With respect to ratio dependence, we have demonstrated the preponderance of evidence in chapters 2, 3, and 4. In this concluding chapter, we focus on an aesthetic consideration. We will try to convince the reader that not using scaling invariance is simply an ugly choice, that certain logical and foundational issues in theoretical ecology will not hang together well without this hypothesis.

6.1 SCALING INVARIANCE AND SYMMETRIES

It is generally accepted that the fundamental model of population dynamics, which describes the growth of a single population in ideal conditions of limitless resources, is the Malthus exponential law. Most ecologists are surprised to learn that joint exponential growth of interacting species expanding (or contracting) at the same rate (a predator and its prey, a parasite and its host) is disallowed by the standard ecological theory, even with limitless resources, and however weak is the interaction. The Lotka-Volterra model predicts that the prey population would grow exponentially in the absence of the predator, but this mathematical solution disappears as soon as the predator population is introduced, and the well-known cyclic solutions appear. In this respect, the model is structurally unstable. Thus, Malthusian growth is reserved to populations of single species in the absence of interactions. This exponential growth is, of course, an idealization even in the case of single species. It cannot hold for long because limiting factors will stop the expansion. That the expansion also has to stop eventually in the case of interacting species, we do not doubt. We question, however, the prevailing model of species interactions because it disallows joint exponential growth even in the ideal conditions of limitless resources and thus sets the single species case in fundamental opposition to that of interacting species (Ginzburg and Colyvan 2004). For our part, we consider that a good model of interacting species must be fundamentally invariant to a proportional change of all abundances in the system.

Let us consider the simplest model of interacting species that intentionally does not include limiting factors:

$$\frac{dN}{dt} = rN - g(\cdot)P$$
$$\frac{dP}{dt} = eg(\cdot)P - qP$$
(6.1)

In the absence of the predator, the prey grows exponentially with rate r. If the predator is present, can both populations grow exponentially? If the argument of the functional response $g(\cdot)$ is the ratio of prey to predator abundances, N/P, the answer is yes. Such a ratio-dependent model allows for joint exponential growth. In order to see this, assume that $N(t) = N_0 \exp(\lambda t)$ and $P(t) = P_0 \exp(\lambda t)$. Then, eqs. (6.1) are equivalent to:

$$\lambda = r - g\left(\frac{N_0}{P_0}\right)\frac{1}{N_0/P_0}$$
$$\lambda = eg\left(\frac{N_0}{P_0}\right) - q \tag{6.2}$$

With a reasonable form of $g(\cdot)$, eqs. (6.2) can be solved simultaneously for the rate λ and the initial ratio N_0/P_0 so that the balanced exponential growth satisfies eqs. (6.1). Quite sensibly, the rate of joint growth λ is smaller than r, the rate of prey growth in isolation. What is true for an interacting pair easily generalizes to a food chain.

For simplicity, we have derived our argument assuming in eqs. (6.1) the linear biomass conversion rule (eq. 1.26). The argument remains valid with a more general conversion rule of the type of eq. (1.24).

Neither the standard prey-dependent models nor the more general predator-dependent models allow for balanced growth. Then, the question is whether allowing for this behavior is a desirable property of the theory of predator-prey interaction. The ideal of exponential growth, even for a single species, requires mixing or elimination of any spatial arrangement of individuals. A cell culture cannot grow faster than quadratically in time if it is restricted to a two-dimensional surface with offspring produced in the vicinity of parents and not mixed. It would be cubic growth if restricted to three dimensions. It is when the spatial arrangement is destroyed, when there is mixing, that all individuals perceive the environment identically as limitless, and that exponential growth emerges. Holt (2009) has defended the same view that movement is fundamental to the exponential assumption. This can be the active movement of organisms or the passive movement of their resources. What is really needed for exponential growth is the obliteration of the spatial structure. The passive movement of cells mixed in a chemostat is a perfect example.

Newton's law of inertia in physics has been used as a metaphor for the Malthus law of population growth. In the same way that Newton's law says how bodies move in the absence of external forces, the Malthus law says how populations grow in the absence of external constraints (Ginzburg 1986; Turchin 2001; Berryman 2002; Ginzburg and Colyvan 2004). Extending this metaphor to include interacting species requires a ratio-dependent description of the interactions. In the language of the metaphor, the standard view is like saying that only a single body can move uniformly in space, but two cannot. In ecology, single species do not really exist. All species have, often unnoticeable, various parasites and commensals. If exponential growth were really restricted to the true single species case, no applications of population dynamics would be possible.

Allowing interacting populations to expand in balanced exponential growth makes the laws of ecology invariant with respect to multiplying interacting abundances by the same constant, so that only ratios matter. This invariance, while certainly approximate, is beautiful because it coordinates the foundational axioms to allow balanced exponential growth, whether it relates to a single species or to an interacting pair. Note that balanced growth is a standard concept in theoretical economics (Cooley 1995). It stands for the exponential growth of the economy with constant ratios between various economic sectors.

Symmetries, which are essentially the same as invariances, are basic to all theories. Even though the most common references to symmetries are in physics, the concept is applicable to any theory, and there is no reason to make an exception for ecology. You do not know what something is until you know what it is not, what transformations or changes of perspective do or do not affect the interaction under study. So the true understanding of the nature of any interaction must include the set of transformations that will not affect it. Examples from physics abound: physical laws are invariant with respect to rotation (no preferred direction in space); the macroscopic laws of classical physics are invariant with respect to time reversal (future and past are indistinguishable). Even though the last one is not true in quantum physics, other more intricate invariances populate modern physics. In fact, they have been developed so far that accepting a full list of symmetries as axioms makes them equivalent to the laws themselves (Wigner 1960). That is, the laws are unique if all the symmetries are true. With all this beautiful development in mind, it is very important to understand that invariances can never be perfect. Richard Feynman gives a very clear explanation of this. He uses the invariance to translation to make the point. Since this invariance is very close in spirit to ratio dependence in ecology, we will start by reconstructing his argument here.

The meaning of translation symmetry is the following: "If you build any kind of apparatus, or do any kind of experiment with some things, and then go and build the same apparatus to do the same kind of experiment with similar things but put them there instead of here, merely translated from one place to another in space, then the same thing will happen in the translated experiment as would have happened in the original experiment" (Feynman 1965, 79).

What is wrong with this logic? Say the experiment involves a pendulum and we move it 50,000 miles to the right. The invariance will not work because you also have to move the Earth 50,000 miles with it. How about the moon? The sun? Soon we will move the whole universe in order to keep our principle acting properly. But this is certainly not the meaning of the original statement. Copying all the universe leads to a trivially identical one, which is of no interest. However, not copying all will not lead to a perfectly identical experimental outcome. The simple conclusion is: our invariance can only be approximate. In practice, for a given problem, if we do not move too far, we can keep the Earth and the moon where they are and the crude invariance will still hold.

Many ecologists believe that physical laws are so perfect that we cannot hope to emulate them. They argue that the high level of biological organization is hopelessly complex and that we try to come up with theories that overlook many important details, not to mention the limited available data. What this argumentation does not appreciate is that the situation is exactly the same in physics: physical laws are not perfectly correct; they are approximations. The excellent book by Nancy Cartwright (1983) titled *How the Laws of Physics Lie* explains this eloquently.

The ratio-dependent invariance is an approximation very similar in spirit to the translation invariance in physics. Since the functional response is a function of the ratio of two abundances, it is thus invariant to joint multiplication. In translation, all coordinates of a physical system get an additive shift. In ratio dependence, it is multiplication instead of addition (with commonly used log-transformed population abundances, it would be additive). Biologically, this means perfect interference of consumers in accessing the food. This cannot be exactly true. In fact, in the limit of low numbers, consumers may not interfere at all and a prey-dependent model will work better than a ratio-dependent one (Abrams and Ginzburg 2000). As was discussed in section 1.6, the functional response may be viewed as prey dependent at the limit of low consumer numbers and ratio dependent at the limit of high consumer numbers (see the models of Ginzburg and Jensen [2008]; Tyutyunov et al. [2008]; Tran [2009]). The question is where on this continuum most natural systems fall, and we have attempted to show in this book (chapters 2, 3, and 4) that they are often at or close to the ratio-dependence end of the continuum. Just like spatial translation in physics, our suggested simple invariance cannot be precise; it has to be approximate. It is violated in extreme conditions, but approximately true in many cases.

Arguments based on invariances are often dismissed by biologists as "physics envy." We are in complete disagreement with this viewpoint. A melody in A minor sounds quite similar after transposition to D minor and can be said to be invariant with respect to such transposition. Invariances are general tools of understanding and as such have applications everywhere: in physics, in music, in ecology, and wherever theories are attempted. Physics as a mature science has much more use of them (and thus more understanding) than younger sciences. However, there is nothing specific in physics for the use of this theoretical tool, which brings clarity and understanding. While always approximate (we will not hear a melody that is transposed too high or too low), invariances help if the range of their validity coincides with the range of our interest. As we have shown in this book, ratio dependence is such an approximate invariance. If the history of physics is any guide, we can expect more elaborate invariances (symmetries) to be discovered in the future. While we do not expect them to be similar to those of physics, we expect them to be similarly illuminating (Colyvan and Ginzburg 2010).

Balanced exponential growth can be viewed as a consequence of the scaling invariance. Or, as we have chosen to present it, the scaling invariance is required if we wish to preserve the possibility of joint exponential growth of an interacting pair. These two statements are logically equivalent or, in mathematical language, "necessary and sufficient conditions." Which to consider the premise and which the consequence is anyone's choice. We have chosen balanced growth as the premise because exponential growth is always treated as a foundation in ecological textbooks. Tautologies in the foundation of theories are much more frequent than one may think. What follows from what may be irrelevant within a theory but one of the two may allow for useful generalizations to a larger theory.

A potential generalization of ratio dependence named Malthusian invariance was proposed in Ginzburg and Colyvan (2004, 95–100). In order to understand it, consider a situation in which two populations, prey and predator, equilibrate at some ratio, say, one predator per 100 prey. The absolute densities are stationary because of a limiting factor controlling the prey population. Now, let us assume that this limiting factor (e.g., space, nutrients, or light) is available in abundance and that the prey population grows, with the predator population growing along with it. If such growth continues, a ratio of abundances for predator and prey (both increasing exponentially because there is no limit) will be established. The question is whether this ratio, with all other things being equal, is different from 1:100. In other words, the question is whether the predator population growth rate will react to the fact that the underlying resource is expanding rather than being constant. We have no doubt that such a reaction will take place with nonexponential expansion of the prey, because in this case the per capita supply of resources will change. The answer is not clear-cut in the case of purely exponential expansion. The

Malthusian invariance principle suggests that the established ratio will not depend on the rate of expansion (or contraction, which might be easier to imagine).

As shown in previous chapters, we have substantial empirical justification for ratio dependence but not for this larger view. It is more on the basis of elegance and simplicity than on direct evidence that we propose it. There is certainly much more experimental and theoretical work to be done to decide whether Malthusian symmetry would become another useful invariance in population dynamic theory.

6.2 KOLMOGOROV'S INSIGHT

In a largely unknown article, Kolmogorov (1936) suggested the possibility of ratio-dependent interaction. The most accessible exposition of the idea is in May's (1974, 86–91) classic book. Attempting to revise the popular Lotka-Volterra model, Kolmogorov considered the properties of the specific growth rates along the rays (kN, kP), where k is an arbitrary positive constant. The ratios of the two abundances are, of course, constant along these rays. In general,

$$\frac{dN}{dt} = NF(N,P)$$
$$\frac{dP}{dt} = PG(N,P)$$
(6.3)

where F and G are the specific growth rates of the prey and the predators respectively. The question Kolmogorov raised was about the relation of $F(kN, kP)$ with $F(N, P)$ and $G(kN, kP)$ with $G(N, P)$ for $k > 1$. In other words, how would prey and predator populations react to a proportional increase in both abundances? His suggested assumption was that there would be a reaction and it would be negative for the prey:

$$F(kN, kP) < F(N, P) \qquad (6.4)$$

He was not as sure in the case of the predator, and included the possibility of equality in the predator's reaction to the proportional change:

$$G(kN, kP) \geq G(N, P) \qquad (6.5)$$

Andrey Nikolaevich Kolmogorov (b. Tambov 1903; d. Moscow 1987). One of the most eminent mathematicians of the twentieth century, he made contributions to almost all areas of mathematics, among them probability theory, topology, intuitionistic logic, turbulence, classical mechanics, and computational complexity. As he became aware of Volterra's work, he published (in Italian) an article in which he generalized this predator-prey theory to much broader conditions (Kolmogorov 1936). See Wikipedia for additional historical information. Drawing by Ksenia Golubkov.

Kolmogorov made a number of other assumptions about the functions F and G, which we do not review here. The essence of the article was to suggest limit cycle oscillations as opposed to Lotka-Volterra conservative cycles. Here, we are interested only in the main assumptions (6.4) and (6.5).

Let us review these assumptions from our point of view. A realistic prey dynamics model always includes density dependence, which means that prey abundance is limited in growth even in the absence of predators. Our suggested ratio-dependent interaction term is invariant to scale change, so the overall response of the prey growth rate F satisfies Kolmogorov's condition (6.4). The response of our predator growth rate G is neutral, which is an allowed borderline case of the Kolmogorov assumption (6.5).

If we include the generalizations suggested in chapter 1, that is, the gradual predator dependence tending toward prey dependence for low P and toward ratio dependence for high P, the inequality for the predator function G becomes strict and our generalized models fit the more general case of Kolmogorov's assumptions completely.

By attracting attention to behavior along the rays of proportional change of both abundances, Kolmogorov's work contained the nucleus of the ratio-dependent theory. At least for one of us, this article was quite influential: LRG remembers meeting Kolmogorov in person once in St. Petersburg to discuss this specific essay. He was then in his 20s and Kolmogorov in his 70s.

6.3 AKÇAKAYA'S RATIO-DEPENDENT MODEL FOR LYNX-HARE CYCLING

Akçakaya's (1992) ratio-dependent model of the most famous data set for Canadian lynx and hare cycles is in many ways the best one available for any specialized predator-prey case study. Not only does it have the smallest number of parameters in comparison to others, but also every parameter is derived from independent data—not from the overall fitting. The only unobserved parameter of the model for which one has no direct information is the size of hare refuges.

The presence of refuges is vital to the outcome of the model. In their absence, model predators consume prey toward extinction and then die themselves, just as in the classical Gause (1934a, 1934b) laboratory cultures. Refuges allow a portion of prey to survive and then, when predators decline sufficiently, to ignite the cycle of increase followed by decline. The period of the cycle (around 10 years) is shown to be largely insensitive to the exact size of the refuges, thus confirming the robustness of the model. This insensitivity is quite understandable. The growth of the nearly unconstrained prey is explosive, almost exponential, and the timing of this explosion (tamed thereafter by the predator increasing along) is mostly controlled by various consumption and reproduction rates, not by the size of the initial inoculum. A similar insensitivity of the period to the prey addition frequency was observed by Gause (1934a, 1934b) in his *Paramecium-Didinium* experiments. The reader is referred to the original publication (Akçakaya 1992), which is based on his PhD thesis, defended at Stony Brook University in 1989.

Table 6.1 compares the Akçakaya model to two published models for cycling mammals. It lists the total number of model parameters along with two other important characteristics: the number of unsupported parameters (i.e., those for which there are no substantial data) and the number of parameters describing the theory's goal. This goal may be to explain the period of observed oscillations: this would be one parameter. If one wishes to describe the amplitude, this would add a parameter. If two species oscillate with different amplitudes, there would be three parameters, with a fourth for the time lag between them. If the cycle asymmetry is to be explained, this adds a fifth parameter. A good theory does not use more parameters than its goal. In other words, the parametric dimension

Table 6.1. OVERFITTING IN ASTRONOMICAL AND IN ECOLOGICAL MODELS

	Total Number of Parameters	Number of Unsupported Parameters	Number of Parameters Describing the Theory Goal	Degree of Overfitting
Theories explaining the motions of planets in the solar system				
Newton (1687)	5	0	5	0
Ptolemy (ca. 150 A.D.)	7	7	5	2
Theories explaining predator-prey mammal cycles				
Akçakaya (1992)	5	1	5	0
Turchin and Batzli (2001)	7	4	3	4
King and Schaffer (2001)	11	3	5	6

Modified from Ginzburg and Jensen (2004), with Ptolemy's model corrected after Fitzpatrick (2008).

of the theory must be as close as possible to the dimension of what it tries to describe.

The last column of table 6.1 shows the degree of overfitting, that is, the excess dimension of the theory over that of the goal. Notably, the Newtonian model does not overfit the goal and all its parameters are supported, while none of the Ptolemaic model is.[1] The situation in predator-prey models is more nuanced. Akçakaya's model has one unsupported parameter (the refuge size), and the other two models have three or four. Akçakaya's model has no overfitting, while the others' degree is four to six, more than the Ptolemaic model. Comparison is therefore in favor of the ratio-dependent Akçakaya model.

An analysis of high-quality data on lynx and hare (Hone et al. 2007) fully supported the ratio-dependent model. With the lynx abundance estimated

---

1. The Newtonian description of planetary ellipses uses the mass of the planet, two initial coordinates, and two components of the initial velocity, that is, a total of five. The mass of the sun and the gravitational constant are two more, but they are not counted because they do not change from planet to planet. An ellipse needs five parameters for a full description: two pairs of coordinates for the two foci plus one length parameter. Therefore, the Newtonian model is not overfitted at all. The traditional characterization of the Ptolemaic system was that it required an enormous number of parameters, with up to 80 nested epicycles. A modern reappraisal shows that Ptolemy's original model was much more parsimonious than that (Fitzpatrick 2008). By our count, the number of parameters for each planet is at least seven, all unsupported, that is, fitted to the observations with no independent measurement.

from a large data set of snow tracks (1987–2004), the lynx annual growth rate is remarkably well predicted by the ratio of hare-to-lynx densities. The specific model is logarithmic (Gompertz-like), not exactly the Akçakaya model but very similar to it. A highly correlated but smaller lynx-count data set (1987–1995) suggests an intermediate predator-dependent relationship, which is somewhat away from pure ratio dependence. For us, the coauthorship of Charles Krebs in the Hone et al. (2007) article is also quite significant, since his contribution to the lynx-hare story unquestionably exceeds that of most other ecologists. So, the truth, if not exactly at the extreme of ratio dependence, is very close to it, as we have claimed all along since 1989.

6.4 THE LIMIT MYTH

There are a number of well-entrenched misconceptions about quantitative natural laws and the role they play in science. The first is that they must be perfect and without exception. The second is that laws should make precise predictions. As we explained in section 6.1, such views are erroneous (see also Ginzburg and Colyvan 2004, section 2.3). In fact, even in physics, laws always appeal to idealizations like a two-body problem (ignoring others) or the absence of friction. The situations with more bodies or with friction are conveniently relegated to engineering or to meteorology, and physics keeps only the idealized part.

These idealizations were called *limit myths* by the philosopher Quine (1960), an expression that caught on because of the very clear image that it suggests. Thus, fundamental laws tell us what happens in the idealized limit and this is apparently very useful. They roughly describe the general disposition of physical and biological systems toward certain behaviors even though real systems will always be somewhat different. Ratio dependence is such a limit myth or, as we have also called it, the "null model of species interaction." Ratio-dependent consumption is a structural suggestion, purely correct only in the extreme, with perfectly even resource sharing. This limit myth gives theory a structure. We can at least start our thinking with this structure and then bend away from it toward more realistic descriptions.

Even though physical analogies have been used in ecology beginning with Lotka and Volterra (Colyvan and Ginzburg 2010), there is one way in which the difference between the two fields is fundamental. Physics deals with the bedrock of all sciences; there is nothing underlying it to serve as a structure from which to deduce basic physical laws. Physical laws are the primary axioms of further deductive work, but in themselves they do not require underlying explanations. Ecology, on the other hand, is highly derived. The time and space scales of biological processes are comparable to human reaction times and perceived spaces. Thus, if physical theory, in the words of Weinberg (1992), is "made out of porcelain," ecological theory is made out of rubber. The meaning of this image is that any small deviation from the core limit myths of physics is disallowed, the rigid porcelain breaks into unconnected pieces. Twisting the ecological theory is much more allowable. Our limit myths are not as precise as they are in physics. In view of this softness, we value, in particular, the special pieces of theory that we view as beautiful or elegant precisely because they are not subject to arbitrary twisting without the violation of a principle. Malthusian or exponential growth is such a structure, which describes "what happens when nothing happens," the null model of ecology. Ecology is all about the deviations from this background—not unlike physics, which is about deviations from uniform motion (Ginzburg and Colyvan 2004). As we have shown, if predation must be compatible with joint exponential expansion, then the ratio-dependent relationship is the necessary limit myth.

It is not an absolute necessity to build population dynamic theory on the foundation of exponential growth. Apparently, most ecologists and all textbooks have decided to accept this view. Given this, we find it completely inconsistent to accept at the same time the Lotka-Volterra axioms (or more general prey-dependent axioms), which disallow joint exponential growth of interacting populations even when resources are unlimited. In our view, communities must be expected to expand exponentially in the presence of unlimited resources. Of course, limiting factors ultimately stop this expansion just as they do for a single species. With our view, it is the limiting resources that stop the joint expansion of the interacting populations; it is not directly due to the interactions themselves. This partitioning of the causes is a major simplification that traditional theory implies only in the case of a single species. In the same way that the Malthus model is accepted as the null model of single species dynamics, we suggest that it is logical that ratio dependence should be viewed as the null model of species interaction. We sketch in appendix 6.A the structure of the ecology textbook that may evolve if our proposed revision is accepted.

In most places in the world today, we face pressing ecological problems due to habitat and species losses, biological invasions, and climate change. Attempting to forecast the consequences of these changes requires a robust theory of how species interact. The existing food web theory, which is essentially built on the Lotka-Volterra null model, is unreliable. Thus, the goal of explaining ecosystem responses to species gains and losses is not well served by the standard prey-dependent view (see a review of the problem by Wardle et al. [2011]). A complementary

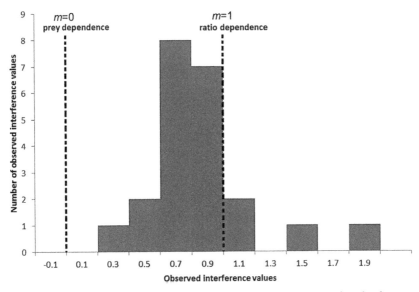

Figure 6.1. Empirical distribution of all available 22 interference parameters m based on bivariate data. See chapter 2 for details.

long-standing unresolved problem is how to explain the causes of species richness patterns (Wiens 2011). Both issues can be resolved only by a food web theory that rests on a representation of elemental interactions that has been validated both by empirical facts and by theoretical arguments. In section 3.6, we briefly pointed out why donor-controlled webs can respond to this challenge. This is only a conjecture at the moment, but we believe it is a credible proposal for further investigation.

In this book, we have presented three kinds of evidence to suggest that an alternative predator-prey interaction theory can be built on the ratio-dependent foundation. The first kind of evidence is based on direct functional response measurements (chapter 2) and figure 6.1 summarizes all available unbiased interference estimates. The second kind of evidence is the indirect community-wide observations of responses to increased primary production (chapter 3). The third kind of evidence rests on mechanistic approaches, suggesting that widespread spatial heterogeneity of the prey distribution is the main reason for predator dependence (chapter 4). If these arguments have raised sufficient doubt in the reader's mind to ponder the idea of revising the basic null model of predation theory, we have achieved our goal.

Appendixes

3.A FOOD CHAIN RESPONSES TO INCREASED PRIMARY PRODUCTION

This appendix brings mathematical proofs to the patterns shown in table 3.1. We will treat the prey-dependent four-level case and the ratio-dependent three-level case. The method can easily be adapted to longer and shorter chains. For simplicity, we assume that primary production is represented by an exogenous input F to the lowermost level N, and that only the uppermost level suffers nonpredatory mortality q. The variables $P_1, P_2, \ldots$ are the sequential consumer levels; the functions $g_1, g_2, \ldots$ are the functional responses; $e_1, e_2, \ldots$ are the conversion efficiencies. We assume that all functional responses $g(\cdot)$ start from zero, are monotonic and concave (i.e., as in figure 1.2), and all parameters are positive. The concavity arises from factors (such as handling time) that result in a diminishing returns effect.

3.A.1 Prey-Dependent Four-Level Food Chain

The dynamic model is

$$\frac{dN}{dt} = F - g_1(N)P_1 \tag{3.A1}$$

$$\frac{dP_1}{dt} = e_1 g_1(N)P_1 - g_2(P_1)P_2 \tag{3.A2}$$

$$\frac{dP_2}{dt} = e_2 g_2(P_1)P_2 - g_3(P_2)P_3 \tag{3.A3}$$

$$\frac{dP_3}{dt} = e_3 g_3(P_2)P_3 - qP_3 \tag{3.A4}$$

To find the population equilibria N^*, P_1^*, ..., we set all four equations equal to zero. We get from the last equation (3.A4), with g_3^{-1} being the inverse function of g_3:

$$P_2^* = g_3^{-1}\left(\frac{q}{e_3}\right) = \text{const.} \qquad (3.A5)$$

Thus, the next-to-top level P_2^* does not vary with primary production F.

Making now sequential substitutions of eq. (3.A1) into (3.A2), then into (3.A3), then into (3.A4), we obtain

$$e_3 e_2 e_1 F - q P_3^* = 0 \qquad (3.A6)$$

that is,

$$P_3^* = \frac{e_3 e_2 e_1}{q} F \qquad (3.A7)$$

which means that the top level P_3^* increases in direct proportion with primary production F.

With the substitution of eq. (3.A1) into (3.A2), we obtain:

$$e_1 F - g_2(P_1^*) P_2^* = 0 \qquad (3.A8)$$

that is,

$$P_1^* = g_2^{-1}\left(\frac{e_1}{P_2^*} F\right) \qquad (3.A9)$$

Remembering that P_2^* is a constant (given by eq. 3.A5) and that g_2 increases concavely, eq. (3.A9) means that P_1^* increases convexly with primary production F, that is, faster than linearly. We now denote it as $P_1^*(F)$ to emphasize that it varies with F.

Finally, the biomass of primary producers N^* can be obtained from eq. (3.A1) as:

$$g_1(N^*) = \frac{F}{P_1^*(F)} \qquad (3.A10)$$

Since we have just shown that $P_1^*(F)$ increases faster than linearly with F, the right-hand side of (3.A10) decreases with F. Since $g_1(\cdot)$ is increasing and invertible, N^*, given by

$$N^* = g_1^{-1}\left(\frac{F}{P_1^*(F)}\right) \qquad (3.\text{A}11)$$

decreases with F, just like the right-hand side of eq. (3.A10).

3.A.2 Ratio-Dependent Three-Level Food Chain

$$\frac{dN}{dt} = F - g_1\left(\frac{N}{P_1}\right)P_1 \qquad (3.\text{A}12)$$

$$\frac{dP_1}{dt} = e_1 g_1\left(\frac{N}{P_1}\right)P_1 - g_2\left(\frac{P_1}{P_2}\right)P_2 \qquad (3.\text{A}13)$$

$$\frac{dP_2}{dt} = e_2 g_2\left(\frac{P_1}{P_2}\right)P_2 - qP_2 \qquad (3.\text{A}14)$$

To find the population equilibria N^*, P_1^*, P_2^*, we set all three equations equal to zero. We get from eq. (3.A14):

$$g_2\left(\frac{P_1^*}{P_2^*}\right) = \frac{q}{e_2}, \text{ hence } \frac{P_1^*}{P_2^*} = g_2^{-1}\left(\frac{q}{e_2}\right) \equiv A = \text{const.} \qquad (3.\text{A}15)$$

Equation (3.A13) gives

$$g_1\left(\frac{N^*}{P_1^*}\right) = \frac{1}{e_1}g_2\left(\frac{P_1^*}{P_2^*}\right)\frac{P_2^*}{P_1^*} \qquad (3.\text{A}16)$$

which gives, with eq. (3.A15),

$$F - \frac{q}{e_1 e_2 A}P_1^* = 0 \qquad (3.\text{A}17)$$

Finally, eq. (3.A12) gives, with (3.A16)

$$F - \frac{q}{e_1 e_2 A}P_1^* = 0 \qquad (3.\text{A}18)$$

Therefore, all three equilibrium population sizes are proportional to F:

$$P_1^* = \frac{e_1 e_2 A}{q} F \qquad (3.A19)$$

$$P_2^* = \frac{P_1^*}{A} = \frac{e_1 e_2}{q} F \qquad (3.A20)$$

$$N^* = B P_1^* = \frac{e_1 e_2 AB}{q} F \qquad (3.A21)$$

3.B CASCADING RESPONSE IN THE RATIO-DEPENDENT MODEL

This appendix demonstrates that in a ratio-dependent trophic chain model, a three-level system has a higher equilibrium biomass on the first trophic level than a two-level system with identical parameters and the top predator eliminated. The calculations are due to H. R. Akçakaya (personal communication). The three-level system is

$$\frac{dP}{dt} = f\left(\frac{N}{P}\right) P - q_P P - g\left(\frac{P}{Z}\right) Z$$

$$\frac{dZ}{dt} = e_Z g\left(\frac{P}{Z}\right) Z - q_Z Z - h\left(\frac{Z}{F}\right) F \qquad (3.B1)$$

$$\frac{dF}{dt} = e_F h\left(\frac{Z}{F}\right) F - q_F F$$

where N is the (constant) nutrient; P, Z, and F are phytoplankton, zooplankton, and fish population densities, respectively; f is the phytoplankton productivity; g and h are the two functional responses; e_Z and e_F are two conversion efficiencies; and q_P, q_Z, and q_F are the three intrinsic death rates. The two-level system is the same without fish: the third equation and the last term of the second equation are both absent. We assume that all functional responses are monotonic and concave (i.e., with negative second derivatives, as in figure 1.2) and all parameters are positive.

To find the population equilibria P^*, Z^*, and F^*, we set all three equations equal to zero. We get from the third equation:

$$\frac{Z^*}{F^*} = h^{-1}\left(\frac{q_F}{e_F}\right) \qquad (3.B2)$$

where h^{-1} is the inverse function of h. Substituting eq. (3.B2) into the second equation of (3.B1), we get

$$\frac{P^*}{Z^*} = g^{-1}(A+B) \tag{3.B3}$$

where g^{-1} is the inverse function of g, and

$$A = \frac{q_Z}{e_Z} \tag{3.B4}$$

$$B = \frac{q_F}{e_Z e_F h^{-1}\left(\frac{q_F}{e_F}\right)} \tag{3.B5}$$

with $A > 0$, $B > 0$. Substituting eq. (3.B3) into the first equation of (3.B1) and rearranging, we obtain the equilibrium phytoplankton biomass in the three-level system as

$$P_3^* = \frac{N}{f^{-1}\left(q_P + \frac{A+B}{g^{-1}(A+B)}\right)} \tag{3.B6}$$

where f^{-1} is the inverse function of f.

For the two-level system, similar calculations give

$$P_2^* = \frac{N}{f^{-1}\left(q_P + \frac{A}{g^{-1}(A)}\right)} \tag{3.B7}$$

Since g is concave (has a negative second derivative), g^{-1} is convex (has a positive second derivative), and thus

$$\frac{A+B}{g^{-1}(A+B)} < \frac{A}{g^{-1}(A)} \tag{3.B8}$$

and since f^{-1} is monotonically increasing, (3.B8) implies that

$$f^{-1}\left(q_P + \frac{A+B}{g^{-1}(A+B)}\right) < f^{-1}\left(q_P + \frac{A}{g^{-1}(A)}\right) \tag{3.B9}$$

Applying eq. (3.B9) to (3.B6) and (3.B7), we see that $P_3^* > P_2^*$. This means that, in the presence of fish, phytoplankton are more abundant than in the same system with fish removed.

6.A How a Revised Ecology Textbook Could Look

We have attempted to convince the reader that there is a strong inconsistency in the foundation of species interaction theory, as it is now presented in all ecology textbooks: Lotka-Volterra–derived predation theory is incompatible with the universally accepted exponential law for single population dynamics. We sketch here the outline that we would follow if we were writing a new textbook.

We start with exponential growth, classically. We follow with the logistic equation as the simplest expression of equilibration due to limiting factors. We make a small but significant change here. The carrying capacity K can be viewed as proportional to a limiting resource S, $K = \gamma S$, and the logistic equation can be rewritten in the equivalent form:

$$\frac{dN}{dt} = rN\left(1 - \frac{1}{\gamma \cdot S/N}\right) \tag{6.A1}$$

Therefore, the fundamental assumption of the logistic equation is that the specific growth rate of a population is a function that increases asymptotically with the ratio S/P, that is, with per capita resource: this model formalizes the view that population dynamics are essentially ratio dependent. The population growth rate is a hyperbolic function of the ratio S/N, saturating at r when S/N is large. It is clear that other growth functions of S/N are thinkable. So, in general,

$$\frac{dN}{dt} = rNf\left(\frac{S}{N}\right) \tag{6.A2}$$

and the logistic equation is just one simple special case. All of this assumes that the resource S is fixed.

Now, when the resource is another population, with its own dynamics, we move to introduce the simplest consistent predation model, that is, the ratio-dependent model. An important corollary here is that joint exponential growth is possible under this model of interaction. A limiting factor must be introduced to stop the infinite expansion, just as in the case of a single species. This is not a minor difference, and we have provided plenty of evidence in this book to support the scale-invariant view as a better initial simplification.

Of course, pure ratio dependence is a limit myth; there are in reality degrees of interference. These degrees are commonly close enough to the limit to make the simplification useful. However, the use of the gradual interference hypothesis will produce prey dependence as a limiting case at low densities when interference is immaterial. The resulting framework embraces the full spectrum of interaction contexts, from the rarefied setting of Monod's chemostats to the common cases of nature.

The Lotka-Volterra equations, after nearly a century of history, are a well-entrenched model of ecology. However, these equations have never been used for anything practical. They could not be used because they commonly contradict the evidence, as we have abundantly shown in this book. One of the reasons for their appeal is the explanation of cycling. Our explanation of cyclicity is different: we accept the choice of predator-prey causes and single-species (maternal effect) causes of cycles as an unresolved issue. There are pros and cons to both views. The new textbook would cover both options within the ratio-dependent framework (see section 5.4). We would then follow with the consequences of predation theory on community ecology: the cascading response and enrichment response of food chains, stability theory, management issues, and so on.

REFERENCES

Abrams, P. A., 1994. The fallacies of ratio-dependent predation. *Ecology*, 75:1842–1850.
Abrams, P. A., and Ginzburg, L. R., 2000. The nature of predation: prey-dependent, ratio-dependent or neither? *Trends in Ecology and Evolution*, 15:337–341.
Abrams, P. A., and Roth, J., 1994. The responses of unstable food chains to enrichment. *Evolutionary Ecology*, 8:150–171.
Abrams, P. A., and Walters, C. J., 1996. Invulnerable prey and the paradox of enrichment. *Ecology*, 77:1125–1133.
Adams, J. R., Vucetich, L. M., Hedrick, P. W., Peterson, R. O., and Vucetich, J. A., 2011. Genomic sweep and potential genetic rescue during limiting environmental conditions in an isolated wolf population. *Proceedings of the Royal Society B*, 278:3336–3344.
Akçakaya, H. R., 1992. Population cycles of mammals: evidence for a ratio-dependent predation hypothesis. *Ecological Monographs*, 62:119–142.
Akçakaya, H. R., Arditi, R., and Ginzburg, L. R., 1995. Ratio-dependent predation: an abstraction that works. *Ecology*, 76:995–1004.
Anderson, P. W., 1972. More is different: broken symmetry and nature of hierarchical structure of science. *Science*, 177:393–396.
Anderson, R. M., and May, R. M., 1991. *Infectious Diseases of Humans: Dynamics and Control*. Oxford: Oxford University Press.
Andrewartha, H. G., and Birch, L. C., 1954. *The Distribution and Abundance of Animals*. Chicago: University of Chicago Press.
Arditi, R., 1983. A unified model of the functional response of predators and parasitoids. *Journal of Animal Ecology*, 52:293–303.
Arditi, R., Abillon, J. M., and Vieira da Silva, J., 1977. The effect of a time-delay in a predator-prey model. *Mathematical Biosciences*, 33:107–120.
Arditi, R., Abillon, J. M., and Vieira da Silva, J., 1978. A predator-prey model with satiation and intraspecific competition. *Ecological Modelling*, 5:173–191.
Arditi, R., and Akçakaya, H. R., 1990. Underestimation of mutual interference of predators. *Oecologia*, 83:358–361.
Arditi, R., and Berryman, A. A., 1991. The biological control paradox. *Trends in Ecology and Evolution*, 6:32–32.
Arditi, R., Callois, J. M., Tyutyunov, Y., and Jost, C., 2004. Does mutual interference always stabilize predator-prey dynamics? A comparison of models. *Comptes Rendus Biologies*, 327:1037–1057.
Arditi, R., and Ginzburg, L. R., 1989. Coupling in predator prey dynamics: ratio dependence. *Journal of Theoretical Biology*, 139:311–326.

Arditi, R., Ginzburg, L. R., and Akçakaya, H. R., 1991. Variation in plankton densities among lakes: a case for ratio-dependent predation models. *American Naturalist*, 138:1287–1296.

Arditi, R., Ginzburg, L. R., and Perrin, N., 1992. Scale invariance is a reasonable approximation in predation models—reply to Ruxton and Gurney. *Oikos*, 65:336–337.

Arditi, R., Perrin, N., and Saïah, H., 1991. Functional responses and heterogeneities: an experimental test with cladocerans. *Oikos*, 60:69–75.

Arditi, R., and Saïah, H., 1992. Empirical evidence of the role of heterogeneity in ratio-dependent consumption. *Ecology*, 73:1544–1551.

Arditi, R., Tyutyunov, Y., Morgulis, A., Govorukhin, V., and Senina, I., 2001. Directed movement of predators and the emergence of density dependence in predator-prey models. *Theoretical Population Biology*, 59:207–221.

Arnold, V. I., 1990. *Huygens and Barrow, Newton and Hooke*. Basel, Switzerland: Birkhäuser Verlag.

Arruda, J. A., 1979. A consideration of trophic dynamics in some tallgrass prairie farm ponds. *American Midland Naturalist*, 102:254–262.

Beddington, J. R., 1975. Mutual interference between parasites or predators and its effects on searching efficiency. *Journal of Animal Ecology*, 44:331–340.

Beddington, J. R., Free, C. A., and Lawton, J. H., 1978. Characteristics of successful natural enemies in models of biological control of insect pests. *Nature*, 273:513–519.

Beddington, J. R., Hassell, M. P., and Lawton, J. H., 1976. The components of arthropod predation. II. Predator rate of increase. *Journal of Animal Ecology*, 45:165–185.

Begon, M., Harper, J. L., and Townsend, C. R., 2006. *Ecology: Individuals, Populations and Communities*. Oxford: Blackwell.

Bender, E. A., Case, T. J., and Gilpin, M. E., 1984. Perturbation experiments in community ecology theory and practice. *Ecology*, 65:1–13.

Berezovskaya, F., Karev, G., and Arditi, R., 2001. Parametric analysis of the ratio-dependent predator-prey model. *Journal of Mathematical Biology*, 43:221–246.

Bernstein, C., 1981. Dispersal of *Phytoseiulus persimilis* [Acarina: Phytoseiidae] in response to prey density distribution. DPhil thesis, Oxford University.

Berryman, A. A., 1999. *Principles of Population Dynamics and Their Application*. Cheltenham, UK: Stanley Thornes.

Berryman, A. A., 2002. Population: a central concept for ecology? *Oikos*, 97:439–442.

Bishop, M. J., Kelaher, B. P., Lincoln Smith, M. P., York, P. H., and Booth, D. J., 2006. Ratio-dependent response of a temperate Australian estuarine system to sustained nitrogen loading. *Oecologia*, 149:701–708.

Blaine, T. W. and DeAngelis, D. L., 1997. The interaction of spatial scale and predator-prey functional response. *Ecological Modelling*, 95:319–328.

Bohannan, B. J. M., and Lenski, R. E., 1997. Effect of resource enrichment on a chemostat community of bacteria and bacteriophage. *Ecology*, 78:2303–2315.

Borer, E. T., Seabloom, E. W., Shurin, J. B., Anderson, K. E., Blanchette, C. A., Broitman, B., Cooper, S. D., and Halpern, B. S., 2005. What determines the strength of a trophic cascade? *Ecology*, 86:528–537.

Briand, F., and Cohen, J. E., 1987. Environmental correlates of food chain length. *Science*, 238:956–960.

Bulmer, M. G., 1975. Phase relations in the ten-year cycle. *Journal of Animal Ecology*, 44:609–621.

Burnham, K. P., and Anderson, D. R., 1998. *Model Selection and Inference: A Practical Information Theoretical Approach*. New York: Springer-Verlag.

Cabrera F, M. I., 2011. Deterministic approach to the study of the interaction predator–prey in a chemostat with predator mutual interference: implications for the paradox of enrichment. *Ecological Modelling*, 222:598–605.

Calder, W. A., 1983. An allometric approach to population cycles of mammals. *Journal of Theoretical Biology*, 100:275–282.

Cantrell, R. S., and Cosner, C., 1991. The effects of spatial heterogeneity in population dynamics. *Journal of Mathematical Biology*, 29:315–338.

Carpenter, S. R. (ed.), 1988. *Complex Interactions in Lake Communities*. New York: Springer-Verlag.

Cartwright, N., 1983. *How the Laws of Physics Lie*. New York: Oxford University Press.

Caswell, H. (ed.), 2005. *Food Webs: From Connectivity to Energetics*. Advances in Ecological Research 36. San Diego, CA: Academic Press.

Chant, D. A., and Turnbull, A. L., 1966. Effects of predator and prey densities on interactions between goldfish and *Daphnia pulex* (de Geer). *Canadian Journal of Zoology*, 44:285–289.

Chase, J. M., Leibold, M. A., Downing, A. L., and Shurin, J. B., 2000. The effects of productivity, herbivory, and plant species turnover in grassland food webs. *Ecology*, 81:2485–2497.

Chesson, P. L., and Murdoch, W. W., 1986. Aggregation of risk: relationships among host-parasitoid models. *American Naturalist*, 127:696–715.

Cohen, J. E., Briand, F., and Newman, C. M., 1990. *Community Food Webs: Data and Theory*. Berlin: Springer-Verlag.

Colyvan, M., and Ginzburg, L. R., 2010. Analogical thinking in ecology: looking beyond disciplinary boundaries. *Quarterly Review of Biology*, 85:171–182.

Comins, H. N., Hassell, M. P., and May, R. M., 1992. The spatial dynamics of host parasitoid systems. *Journal of Animal Ecology*, 61:735–748.

Contois, D. E., 1959. Kinetics of bacterial growth: relationship between population density and specific growth rate of continuous cultures. *Journal of General Microbiology*, 21: 40–50.

Cook, K., and Busenberg, S., 1993. *Vertically Transmitted Diseases*. New York: Springer-Verlag.

Cooley, T. F. (ed.), 1995. *Frontiers of Business Cycle Research*. Princeton, NJ: Princeton University Press.

Cosner, C., DeAngelis, D. L., Ault, J. S., and Olson, D. B., 1999. Effects of spatial grouping on the functional response of predators. *Theoretical Population Biology*, 56:65–75.

Davison, A. C., and Hinkley, D. V., 1997. *Bootstrap Methods and Their Application*. Cambridge: Cambridge University Press.

DeAngelis, D. L., Goldstein, R. A., and O'Neill, R. V., 1975. A model for trophic interaction. *Ecology*, 56:881–892.

Deevey, E. S., Jr., 1941. Limnological studies in Connecticut. VI. The quantity and composition of the bottom fauna of thirty-six Connecticut and New York lakes. *Ecological Monographs*, 11:413–455.

DeLong, J., and Vasseur, D., 2011. Mutual interference is common and mostly intermediate in magnitude. *BMC Ecology*, 11:1.

De Ruiter, P. C., Wolters, V., Moore, J. C., and Winemiller, K. O. (eds.), 2005. *Dynamic Food Webs: Multispecies Assemblages, Ecosystem Development and Environmental Change*. Amsterdam: Academic Press.

Edelstein-Keshet, L., Watmough, J., and Grünbaum, D., 1998. Do travelling band solutions describe cohesive swarms? An investigation for migratory locusts. *Journal of Mathematical Biology*, 36:515–549.

Edwards, R. L., 1961. The area of discovery of two insect parasites, *Nasonia vitripennis* (Walker) and *Trichogramma evanescens* Westwood, in an artificial environment. *Canadian Entomologist*, 93:475–481.

Efron, B., and Tibshirani, R. J., 1993. *An Introduction to the Bootstrap*. London: Chapman and Hall.

Elmhagen, B., Ludwig, G., Rushton, S. P., Helle, P., and Linden, H., 2010. Top predators, mesopredators and their prey: interference ecosystems along bioclimatic productivity gradients. *Journal of Animal Ecology*, 79:785–794.

Estes, J. A., Terborgh, J., Brashares, J. S., Power, M. E., Berger, J., Bond, W. J., Carpenter, S. R., Essington, T. E., Holt, R. D., Jackson, J. B. C., Marquis, R. J., Oksanen, L., Oksanen, T., Paine, R. T., Pikitch, E. K., Ripple, W. J., Sandin, S. A., Scheffer, M., Schoener, T. W., Shurin, J. B., Sinclair, A. R. E., Soule, M. E., Virtanen, R. and Wardle, D. A., 2011. Trophic downgrading of planet Earth. *Science*, 333:301–306.

Eveleigh, E. S., and Chant, D. A., 1982. Experimental studies on acarine predator-prey interactions: the effects of predator density on prey consumption, predator searching efficiency, and the functional response to prey density (Acarina: Phytoseiidae). *Canadian Journal of Zoology*, 60:611–629.

Farmelo, G. (ed.), 2002. *It Must Be Beautiful: Great Equations of Modern Science*. London: Granta Books.

Feynman, R. P., 1965. *The Character of Physical Law*. London: British Broadcasting Corporation.

Fitzpatrick, R., 2008. *A Modern Almagest: An Updated Version of Ptolemy's Model of the Solar System*. Available from http://farside.ph.utexas.edu/, Austin, TX.

Flierl, G., Grünbaum, D., Levin, S., and Olson, D., 1999. From individuals to aggregations: the interplay between behavior and physics. *Journal of Theoretical Biology*, 196:397–454.

Free, C. A., Beddington, J. R., and Lawton, J. H., 1977. On the inadequacy of simple models of mutual interference for parasitism and predation. *Journal of Animal Ecology*, 46:543–554.

Fretwell, S. D., 1977. The regulation of plant communities by food chains exploiting them. *Perspectives in Biology and Medicine*, 20:169–185.

Fussmann, G. F., Ellner, S. P., Shertzer, K. W., and Hairston, N. G., 2000. Crossing the Hopf bifurcation in a live predator-prey system. *Science*, 290:1358–1360.

Fussmann, G. F., Weithoff, G., and Yoshida, T., 2005. A direct, experimental test of resource vs. consumer dependence. *Ecology*, 86:2924–2930.

Gatto, M., 1991. Some remarks on models of plankton densities in lakes. *American Naturalist*, 137:264–267.

Gause, G. F., 1934a. Experimental analysis of Vito Volterra's mathematical theory of the struggle for existence. *Science*, 79:16–17.

Gause, G. F., 1934b. *The Struggle for Existence*. Baltimore, MD: Williams and Wilkins.

Gause, G. F., 1935a. *Vérifications expérimentales de la théorie mathématique de la lutte pour la vie*. Paris: Hermann.

Gause, G. F., 1935b. Experimental demonstration of Volterra's periodic oscillations in the numbers of animals. *Journal of Experimental Biology*, 12:44–48.

Genkai-Kato, M., and Yamamura, N., 1999. Unpalatable prey resolves the paradox of enrichment. *Proceedings of the Royal Society of London Series B, Biological Sciences*, 266:1215–1219.

Getz, W. M., 1984. Population dynamics: a per capita resource approach. *Journal of Theoretical Biology*, 108:623–643.

Getz, W. M., 1998. An introspection on the art of modeling in population ecology. *Bioscience*, 48:540–552.

Ginzburg, L. R., 1983. *Theory of Natural Selection and Population Growth*. Menlo Park, CA: Benjamin/Cummings.

Ginzburg, L. R., 1986. The theory of population dynamics: back to first principles. *Journal of Theoretical Biology*, 122:385–399.

Ginzburg, L. R., 1998. Assuming reproduction to be a function of consumption raises doubts about some popular predator-prey models. *Journal of Animal Ecology*, 67:325–327.

Ginzburg, L. R. and Akçakaya, H. R., 1992. Consequences of ratio-dependent predation for steady-state properties of ecosystems. *Ecology*, 73:1536–1543.

Ginzburg, L. R., Burger, O., and Damuth, J., 2010. The May threshold and life-history allometry. *Biology Letters*, 6:850–853.

Ginzburg, L. R., and Colyvan, M., 2004. *Ecological Orbits: How Planets Move and Populations Grow*. New York: Oxford University Press.

Ginzburg, L. R., Goldman, Y. I., and Railkin, A. I., 1971. A mathematical model of interaction between two populations. I. Predator-prey [in Russian]. *Zhurnal Obshchei Biologii*, 32:724–730.

Ginzburg, L. R., and Jensen, C. X. J., 2004. Rules of thumb for judging ecological theories. *Trends in Ecology and Evolution*, 19:121–126.

Ginzburg, L. R., and Jensen, C. X. J., 2008. From controversy to consensus: the indirect interference functional response. *Verhandlungen der Internationalen Vereinigung für Theoretische und Angewandte Limnologie*, 30:297–301.

Ginzburg, L. R., Konovalov, N. Y., and Epelman, G. S., 1974. A mathematical model of interaction between two populations. IV. Theoretical and experimental data [in Russian]. *Zhurnal Obshchei Biologii*, 35:613–619.

Gleeson, S. K., 1994. Density dependence is better than ratio dependence. *Ecology*, 75:1834–1935.

Gobler, C. J., Lonsdale, D. J., and Boyer, G. L., 2005. A review of the causes, effects, and potential management of harmful brown tide blooms caused by *Aureococcus anophagefferens* (Hargraves et Sieburth). *Estuaries and Coasts*, 28:726–749.

Gotelli, N. J., 2008. *A Primer of Ecology*. Sunderland, MA: Sinauer.

Greene, M., 2001. A tool, not a tyrant. *Nature*, 410:875.

Grünbaum, D., and Okubo, A., 1994. Modelling social animal aggregations. In: S. Levin (ed.), *Frontiers in Mathematical Biology, Lecture Notes in Biomathematics, Vol. 100*. New York: Springer-Verlag, pp. 296–325.

Gutierrez, A. P., 1996. *Applied Population Ecology: A Supply-Demand Approach*. New York: Wiley.

Gutierrez, A. P., and Baumgaertner, J. U., 1984. Multitrophic level models of predator-prey energetics: I. Age-specific energetics models—Pea aphid *Acyrthosiphon pisum* (Homoptera: Aphididae) as an example. *Canadian Entomologist*, 116:924–932.

Hairston, N. G., Smith, F. E., and Slobodkin, L. B., 1960. Community structure, population control, and competition. *American Naturalist*, 94:421–425.

Hanski, I., 1992. The functional response of predators: worries about scale. *Trends in Ecology and Evolution*, 6:141–142.

Hanson, J. M., and Leggett, W. C., 1982. Empirical prediction of fish biomass and yield. *Canadian Journal of Fisheries and Aquatic Sciences*, 39:257–263.

Hanson, J. M., and Peters, R. H., 1984. Empirical prediction of crustacean zooplankton biomass and profundal macrobenthos biomass in lakes. *Canadian Journal of Fisheries and Aquatic Sciences*, 41:439–445.

Hansson, S., De Stasio, B. T., Gorokhova, E., and Mohammadian, M. A., 2001. Ratio-dependent functional responses—tests with the zooplanktivore *Mysis mixta*. *Marine Ecology Progress Series*, 216:181–189.

Harmand, J., and Godon, J. J., 2007. Density-dependent kinetics models for a simple description of complex phenomena in macroscopic mass-balance modeling of bioreactors. *Ecological Modelling*, 200:393–402.

Harrison, G. W., 1995. Comparing predator-prey models to Luckinbill's experiment with *Didinium* and *Paramecium*. *Ecology*, 76:357–374.

Hassell, M. P., 1978. *The Dynamics of Arthropod Predator-Prey Systems*. Princeton, NJ: Princeton University Press.

Hassell, M. P., 2000. *The Spatial and Temporal Dynamics of Host-Parasitoid Interactions*. Oxford: Oxford University Press.
Hassell, M. P., and Anderson, R. M., 1989. Predator-prey and host-pathogen interactions. In: J. M. Cherrett (ed.), *Ecological Concepts: The Contribution of Ecology to an Understanding of the Natural World*. Oxford: Blackwell Scientific, pp. 147–196.
Hassell, M. P., Lawton, J. H., and Beddington, J. R., 1976. The components of arthropod predation. I. The prey death-rate. *Journal of Animal Ecology*, 45:135–164.
Hassell, M. P., and May, R. M., 1973. Stability in insect host-parasite models. *Journal of Animal Ecology*, 42:693–726.
Hassell, M. P., and May, R. M., 1974. Aggregation of predators and insect parasites and its effect on stability. *Journal of Animal Ecology*, 43:567–594.
Hassell, M. P., and Varley, G. C., 1969. New inductive population model for insect parasites and its bearing on biological control. *Nature*, 223:1133–1137.
Hastings, A., 1990. Spatial heterogeneity and ecological models. *Ecology*, 71:426–428.
Hauzy, C., Tully, T., Spataro, T., Paul, G., and Arditi, R., 2010. Spatial heterogeneity and functional response: an experiment in microcosms with varying obstacle densities. *Oecologia*, 163:625–636.
Hawkins, B. A., Thomas, M. B., and Hochberg, M. E., 1993. Refuge theory and biological control. *Science*, 262:1429–1432.
Holling, C. S., 1959a. The components of predation as revealed by a study of small-mammal predation of the European pine sawfly. *Canadian Entomologist*, 91:293–320.
Holling, C. S., 1959b. Some characteristics of simple types of predation and parasitism. *Canadian Entomologist*, 91:385–398.
Holt, R. D., 2009. Darwin, Malthus, and movement: a hidden assumption in the demographic foundations of evolution. *Israel Journal of Ecology and Evolution*, 55:189–198.
Holt, R. D., 2011. Natural enemy-victim interactions: do we have a unified theory yet? In: S. M. Scheiner and M. R. Willig (eds.), *The Theory of Ecology*. Chicago: University of Chicago Press, pp. 125–161.
Holyoak, M., 2000. Effects of nutrient enrichment on predator-prey metapopulation dynamics. *Journal of Animal Ecology*, 69:985–997.
Hone, J., Krebs, C., O'Donoghue, M., and Boutin, S., 2007. Evaluation of predator numerical responses. *Wildlife Research*, 34:335–341.
Hutchinson, G. E., 1978. *An Introduction to Population Ecology*. New Haven, CT: Yale University Press.
Inchausti, P., and Ginzburg, L. R., 1998. Small mammal cycles in northern Europe: patterns and evidence for a maternal effect hypothesis. *Journal of Animal Ecology*, 67:180–194.
Inchausti, P., and Ginzburg, L. R., 2009. Maternal effects mechanism of population cycling: a formidable competitor to the traditional predator-prey view. *Philosophical Transactions of the Royal Society B, Biological Sciences*, 364:1117–1124.
Ivlev, V. S., 1961. *Experimental Ecology of the Feeding of Fishes*. New Haven, CT: Yale University Press.
Jansen, V. A. A., 1995. Regulation of predator-prey systems through spatial interactions: a possible solution to the paradox of enrichment. *Oikos*, 74:384–390.
Jansen, V. A. A., and De Roos, A. M., 2000. The role of space in reducing predator-prey cycles. In: U. Dieckmann, R. Law, and J. A. J. Metz (eds.), *The Geometry of Ecological Interactions: Simplifying Spatial Complexity*. Cambridge: Cambridge University Press, pp. 183–201.
Jensen, C. X. J., and Ginzburg, L. R., 2005. Paradoxes or theoretical failures? The jury is still out. *Ecological Modelling*, 188:3–14.
Jensen, C. X. J., Jeschke, J. M., and Ginzburg, L. R., 2007. A direct, experimental test of resource vs. consumer dependence: comment. *Ecology*, 88:1600–1602.

Jones, J. R., and Hoyer, M. V., 1982. Sportfish harvest predicted by summer chlorophyll-*a* concentration in Midwestern lakes and reservoirs. *Transactions of the American Fisheries Society*, 111:176–179.

Jones, T. H., and Hassell, M. P., 1988. Patterns of parasitism by *Trybliographa rapae*, a cynipid parasitoid of the cabbage root fly, under laboratory and field conditions. *Ecological Entomology*, 13:309–317.

Jost, C., 2000. Predator-prey theory: hidden twins in ecology and microbiology. *Oikos*, 90:202–208.

Jost, C., and Arditi, R., 2000. Identifying predator-prey processes from time series. *Theoretical Population Biology*, 57:325–337.

Jost, C., and Arditi, R., 2001. From pattern to process: identifying predator-prey models from time-series data. *Population Ecology*, 43:229–243.

Jost, C., Arino, O., and Arditi, R., 1999. About deterministic extinction in ratio-dependent predator-prey models. *Bulletin of Mathematical Biology*, 61:19–32.

Jost, C., Devulder, G., Vucetich, J. A., Peterson, R. O., and Arditi, R., 2005. The wolves of Isle Royale display scale-invariant satiation and ratio-dependent predation on moose. *Journal of Animal Ecology*, 74:809–816.

Jost, C., and Ellner, S. P., 2000. Testing for predator dependence in predator-prey dynamics: a non-parametric approach. *Proceedings of the Royal Society of London Series B, Biological Sciences*, 267:1611–1620.

Katz, C. H., 1985. A nonequilibrium marine predator-prey interaction. *Ecology*, 66:1426–1438.

Kaunzinger, C. M. K., and Morin, P. J., 1998. Productivity controls food-chain properties in microbial communities. *Nature*, 395:495–497.

Kerfoot, W. C., and DeAngelis, D. L., 1989. Scale-dependent dynamics: zooplankton and the stability of freshwater food webs. *Trends in Ecology and Evolution*, 4:167–171.

Kfir, R., 1983. Functional response to host density by the egg parasite *Trichogramma pretiosum*. *Entomophaga*, 28:345–353.

King, A. A., and Schaffer, W. M., 2001. The geometry of a population cycle: a mechanistic model of snowshoe hare demography. *Ecology*, 82:814–830.

Kingsland, S. E., 1985. *Modeling Nature: Episodes in the History of Population Ecology*. Chicago: University of Chicago Press.

Kolmogorov, A. N., 1936. Sulla teoria di Volterra della lotta per l'esistenza. *Giornale dell'Istituto Italiano degli Attuari*, 7:74–80.

Kratina, P., Vos, M., Bateman, A., and Anholt, B. R., 2009. Functional responses modified by predator density. *Oecologia*, 159:425–433.

Krebs, C. J., 2009. *Ecology: The Experimental Analysis of Distribution and Abundance*. San Francisco: Benjamin Cummings.

Kretzschmar, M., Nisbet, R. M., and McCauley, E., 1993. A predator-prey model for zooplankton grazing on competing algal populations. *Theoretical Population Biology*, 44:32–66.

Krukonis, G., and Schaffer, W. M., 1991. Population cycles in mammals and birds: does periodicity scale with body size. *Journal of Theoretical Biology*, 148:469–493.

Kumar, A., and Tripathi, C. P. M., 1985. Parasitoid-host relationship between *Trioxys (Binodoxys) indicus* Subba Rao & Sharma (Hymenoptera: Aphidiidae) and *Aphis craccivora* Koch (Hemiptera: Aphididae): effect of host plants on the area of discovery of the parasitoid. *Canadian Journal of Zoology*, 63:192–195.

Laughlin, R., 2005. *A Different Universe: Remaking Physics from the Bottom Down*. New York: Basic Books.

Leibold, M. A., 1989. Resource edibility and the effects of predators and productivity on the outcome of trophic interactions. *American Naturalist*, 134:922–949.

Leslie, P. H., 1948. Some further notes on the use of matrices in population mathematics. *Biometrika*, 35:213–245.

Lobry, C., and Harmand, J., 2006. A new hypothesis to explain the coexistence of *n* species in the presence of a single resource. *Comptes Rendus Biologies*, 329:40–46.
Lotka, A. J., 1924. *Elements of Physical Biology*. Baltimore: Williams and Wilkins. Republished as *Elements of Mathematical Biology*, Dover, 1956.
Luck, R. F., 1990. Evaluation of natural enemies for biological control: a behavioral approach. *Trends in Ecology and Evolution*, 5:196–199.
Luckinbill, L. S., 1973. Coexistence in laboratory populations of *Paramecium aurelia* and its predator *Didinium nasutum*. *Ecology*, 54:1320–1327.
Luckinbill, L. S., 1974. Effects of space and enrichment on a predator-prey system. *Ecology*, 55:1142–1147.
Lynch, L. D., Bowers, R. G., Begon, M., and Thompson, D. J., 1998. A dynamic refuge model and population regulation by insect parasitoids. *Journal of Animal Ecology*, 67:270–279.
Marshall, C. T., and Peters, R. H., 1989. General patterns in the seasonal development of chlorophyll-*a* for temperate lakes. *Limnology and Oceanography*, 34:856–867.
May, R. M., 1974. *Stability and Complexity in Model Ecosystems*. Princeton: Princeton University Press.
May, R. M. (ed.), 1976. *Theoretical Ecology: Principles and Applications*. Oxford: Blackwell.
May, R. M., 1978. Host-parasitoid systems in patchy environments: a phenomenological model. *Journal of Animal Ecology*, 47:833–844.
May, R. M., 2002. The best possible time to be alive: the logistic map. In: G. Farmelo (ed.), *It Must Be Beautiful: Great Equations of Modern Science*. London: Granta Books, pp. 212–229.
Maynard Smith, J., 2002. Equations of life: the mathematics of evolution. In: G. Farmelo (ed.), *It Must Be Beautiful: Great Equations of Modern Science*. London: Granta Books, pp. 193–211.
Mazumder, A., 1994. Patterns of algal biomass in dominant odd-link vs. even-link lake ecosystems. *Ecology*, 75:1141–1149.
McCauley, E., and Kalff, J., 1981. Empirical relationships between phytoplankton and zooplankton biomass in lakes. *Canadian Journal of Fisheries and Aquatic Sciences*, 38:458–463.
McCauley, E., and Murdoch, W. W., 1990. Predator-prey dynamics in environments rich and poor in nutrients. *Nature*, 343:455–457.
McCauley, E., Murdoch, W. W., and Watson, S., 1988. Simple models and variation in plankton densities among lakes. *American Naturalist*, 132:383–403.
McCauley, E., Nisbet, R. M., Murdoch, W. W., De Roos, A. M., and Gurney, W. S. C., 1999. Large-amplitude cycles of *Daphnia* and its algal prey in enriched environments. *Nature*, 402:653–656.
McNaughton, S. J., Oesterheld, M., Frank, D. A., and Williams, K. J., 1989. Ecosystem-level patterns of primary productivity and herbivory in terrestrial habitats. *Nature*, 341:142–144.
McQueen, D. J., Post, J. R., and Mills, E. L., 1986. Trophic relationships in freshwater pelagic ecosystems. *Canadian Journal of Fisheries and Aquatic Sciences*, 43:1571–1581.
Mertz, D. B., and Davies, R. B., 1968. Cannibalism of pupal stage by adult flour beetles: an experiment and a stochastic model. *Biometrics*, 24:247–275.
Michalski, J., Poggiale, J. C., Arditi, R., and Auger, P. M., 1997. Macroscopic dynamic effects of migrations in patchy predator-prey systems. *Journal of Theoretical Biology*, 185:459–474.
Mills, E. L., and Schiavone, A., Jr., 1982. Evaluation of fish communities through assessment of zooplankton populations and measures of lake productivity. *North American Journal of Fisheries Management*, 2:14–27.

Mills, N. J., and Lacan, I., 2004. Ratio dependence in the functional response of insect parasitoids: evidence from *Trichogramma minutum* foraging for eggs in small host patches. *Ecological Entomology*, 29:208–216.

Monod, J., 1942. *Recherches sur la croissance des cultures bactériennes*. Paris: Hermann.

Monod, J., 1950. La technique de culture continue, théorie et applications. *Annales de l'Institut Pasteur*, 79:390–410.

Mougi, A., and Kishida, O., 2009. Reciprocal phenotypic plasticity can lead to stable predator-prey interaction. *Journal of Animal Ecology*, 78:1172–1181.

Mounier, J., Monnet, C., Vallaeys, T., Arditi, R., Sarthou, A.-S., Hélias, A., and Irlinger, F., 2008. Microbial interactions within a cheese microbial community. *Applied and Environmental Microbiology*, 74:172–181.

Murdoch, W. W., 1994. Population regulation in theory and practice. *Ecology*, 75:271–287.

Murdoch, W. W., Briggs, C. J., and Nisbet, R. M., 2003. *Consumer-Resource Dynamics*. Princeton, NJ: Princeton University Press.

Murdoch, W. W., Chesson, J., and Chesson, P. L., 1985. Biological control in theory and practice. *American Naturalist*, 125:344–366.

Murdoch, W. W., Nisbet, R. M., McCauley, E., De Roos, A. M., and Gurney, W. S. C., 1998. Plankton abundance and dynamics across nutrient levels: tests of hypotheses. *Ecology*, 79:1339–1356.

Nelson, M. I., and Holder, A., 2009. A fundamental analysis of continuous flow bioreactor models governed by Contois kinetics. II. Reactor cascades. *Chemical Engineering Journal*, 149:406–416.

Nisbet, R. M., De Roos, A. M., Wilson, W. G., and Snyder, R. E., 1998. Discrete consumers, small scale resource heterogeneity, and population stability. *Ecology Letters*, 1:34–37.

Novick, A., and Szilard, L., 1950. Experiments with the chemostat on spontaneous mutations of bacteria. *Proceedings of the National Academy of Sciences, USA*, 36: 708–719.

Oksanen, L., Fretwell, S. D., Arruda, J., and Niemela, P., 1981. Exploitation ecosystems in gradients of primary productivity. *American Naturalist*, 118:240–261.

Okubo, A., Chiang, H. C., and Ebbesmeyer, C. C., 1977. Acceleration field of individual midges, *Anarete pritchardi* (Diptera: Cecidomyiidae), within a swarm. *Canadian Entomologist*, 109:149–156.

Owen-Smith, N., 1990. Demography of a large herbivore, the greater kudu, in relation to rainfall. *Journal of Animal Ecology*, 59:893–913.

Owen-Smith, N., 2002. *Adaptive Herbivore Ecology: From Resources to Populations in Variable Environments*. Cambridge: Cambridge University Press.

Pace, M. L., 1984. Zooplankton community structure, but not biomass, influences the phosphorus-chlorophyll *a* relationship. *Canadian Journal of Fisheries and Aquatic Sciences*, 41:1089–1096.

Parrish, J. K., and Turchin, P., 1997. Individual decisions, traffic rules, and emergent pattern: a Lagrangian analysis. In: J. K. Parrish, W. M. Hamner, and C. T. Prewitt (ed.), *Animal Aggregations: Three-Dimensional Measurement and Modelling*. Cambridge: Cambridge University Press, pp. 126–142.

Parrish, J. K., Viscido, S. V., and Grünbaum, D., 2002. Self-organized fish schools: an examination of emergent properties. *Biological Bulletin*, 202:296–305.

Pascual, M., and Levin, S. A., 1999. Spatial scaling in a benthic population model with density-dependent disturbance. *Theoretical Population Biology*, 56:106–122.

Persson, L., Andersson, G., Hamrin, S. F., and Johansson, L., 1988. Predator regulation and primary productivity along the productivity gradient of temperate lake ecosystems. In: S. R. Carpenter (ed.), *Complex Interactions in Lake Communities*. Berlin: Springer-Verlag, pp. 45–65.

Persson, L., Johansson, L., Andersson, G., Diehl, S., and Hamrin, S. F., 1993. Density-dependent interactions in lake ecosystems: whole-lake perturbation experiments. *Oikos*, 66:193–208.

Petrovskii, S., Li, B. L., and Malchow, H., 2004. Transition to spatiotemporal chaos can resolve the paradox of enrichment. *Ecological Complexity*, 1:37–47.

Phillips, O. M., 1974. Equilibrium and stability of simple marine biological systems. 2. Herbivores. *Archiv für Hydrobiologie*, 73:310–333.

Pimm, S. L., 1982. *Food Webs*. London: Chapman and Hall.

Pimm, S. L., 1991. *The Balance of Nature? Ecological Issues in the Conservation of Species and Communities*. Chicago: University of Chicago Press.

Poggiale, J. C., Michalski, J., and Arditi, R., 1998. Emergence of donor control in patchy predator-prey systems. *Bulletin of Mathematical Biology*, 60:1149–1166.

Polis, G. A., and Winemiller, K. O. (eds.), 1996. *Food Webs: Integration of Patterns and Dynamics*. New York: Chapman and Hall.

Ponsard, S., and Arditi, R., 2000. What can stable isotopes ($\delta^{15}N$ and $\delta^{13}C$) tell about the food web of soil macro-invertebrates? *Ecology*, 81:852–864.

Ponsard, S., and Arditi, R., 2001. Detecting omnivory with $\delta^{15}N$. *Trends in Ecology and Evolution*, 16:20–21.

Ponsard, S., Arditi, R., and Jost, C., 2000. Assessing top-down and bottom-up control in a litter-based soil macroinvertebrate food chain. *Oikos*, 89:524–540.

Post, D. M., Pace, M. L., and Hairston, N. G., 2000. Ecosystem size determines food-chain length in lakes. *Nature*, 405:1047–1049.

Prepas, E. E., and Trew, D. O., 1983. Evaluation of the phosphorus-chlorophyll relationship for lakes off the Precambrian Shield in Western Canada. *Canadian Journal of Fisheries and Aquatic Sciences*, 40:27–35.

Prigogine, I., 1997. *The End of Certainty: Time, Chaos and the Laws of Nature*. New York: Free Press.

Quine, W. V., 1960. *Word and Object*. Cambridge, MA: MIT Press.

Reeve, J. D., 1997. Predation and bark beetle dynamics. *Oecologia*, 112:48–54.

Ricklefs, R. E., 1979. *Ecology*. New York: Chiron.

Rogers, D., 1972. Random search and insect population models. *Journal of Animal Ecology*, 41:369–383.

Rosenzweig, M. L., 1969. Why the prey curve has a hump. *American Naturalist*, 103:81–87.

Rosenzweig, M. L., 1971. Paradox of enrichment: destabilization of exploitation ecosystems in ecological time. *Science*, 171:385–387.

Rosenzweig, M. L., 1973. Evolution of the predator isocline. *Evolution*, 27:84–94.

Rosenzweig, M. L., 1977. Aspects of biological exploitation. *Quarterly Review of Biology*, 52:371–380.

Rosenzweig, M. L., and MacArthur, R. H., 1963. Graphical representation and stability conditions of predator-prey interactions. *American Naturalist*, 97:209–223.

Royama, T., 1971. A comparative study of models for predation and parasitism. *Researches on Population Ecology*, S1:1–90.

Ruxton, G. D., and Gurney, W. S. C., 1992. The interpretation of tests for ratio dependence. *Oikos*, 65:334–335.

Sapoukhina, N., Tyutyunov, Y., and Arditi, R., 2003. The role of prey taxis in biological control: a spatial theoretical model. *American Naturalist*, 162:61–76.

Sapoukhina, N. Y., 2002. Modeling spatial dynamics of trophic communities with application to biological control. PhD thesis, Institut National Agronomique Paris-Grignon, Paris.

Schenk, D., and Bacher, S., 2002. Functional response of a generalist insect predator to one of its prey species in the field. *Journal of Animal Ecology*, 71:524–531.

Schenk, D., Bersier, L. F., and Bacher, S., 2005. An experimental test of the nature of predation: neither prey- nor ratio-dependent. *Journal of Animal Ecology*, 74:86–91.

Shurin, J. B., Borer, E. T., Seabloom, E. W., Anderson, K., Blanchette, C. A., Broitman, B., Cooper, S. D., and Halpern, B. S., 2002. A cross-ecosystem comparison of the strength of trophic cascades. *Ecology Letters*, 5:785–791.

Shurin, J. B., and Seabloom, E. W., 2005. The strength of trophic cascades across ecosystems: predictions from allometry and energetics. *Journal of Animal Ecology*, 74:1029–1038.

Skalski, G. T., and Gilliam, J. F., 2001. Functional responses with predator interference: viable alternatives to the Holling Type II model. *Ecology*, 82:3083–3092.

Sommer, U., and Sommer, F., 2006. Cladocerans versus copepods: the cause of contrasting top-down controls on freshwater and marine phytoplankton. *Oecologia*, 147:183–194.

Spataro, T., Bacher, S., Bersier, L. F., and Arditi, R., 2011. Ratio-dependent predation in a field experiment with wasps. Unpublished manuscript.

Stibor, H., Vadstein, O., Diehl, S., Gelzleichter, A., Hansen, T., Hantzsche, F., Katechakis, A., Lippert, B., Loseth, K., Peters, C., Roederer, W., Sandow, M., Sundt-Hansen, L., and Olsen, Y., 2004. Copepods act as a switch between alternative trophic cascades in marine pelagic food webs. *Ecology Letters*, 7:321–328.

Stockner, I. G., and Shortreed, K. S., 1985. Whole-lake fertilization experiments in coastal British Columbia lakes: empirical relationships between nutrient inputs and phytoplankton biomass and production. *Canadian Journal of Fisheries and Aquatic Sciences*, 42:649–658.

Sutherland, W. J., 1996. *From Individual Behaviour to Population Ecology*. Oxford: Oxford University Press.

Thompson, W. R., 1924. La théorie mathématique de l'action des parasites entomophages et le facteur du hasard. *Annales de la Faculté des Sciences de Marseille*, 2:69–89.

Thurber, J. M., and Peterson, R. O., 1993. Effects of population-density and pack size on the foraging ecology of gray wolves. *Journal of Mammalogy*, 74:879–889.

Tilman, D., 1994. Competition and biodiversity in spatially structured habitats. *Ecology*, 75:2–16.

Tran, J. K., 2009. Discrete reproduction leads to an intermediate predator-prey functional response. *Verhandlungen der Internationalen Vereinigung für Theoretische und Angewandte Limnologie*, 30:1310–1312.

Turchin, P., 2001. Does population ecology have general laws? *Oikos*, 94:17–26.

Turchin, P., 2003. *Complex Population Dynamics: A Theoretical/Empirical Synthesis*. Princeton, NJ: Princeton University Press.

Turchin, P., and Batzli, G. O., 2001. Availability of food and the population dynamics of arvicoline rodents. *Ecology*, 82:1521–1534.

Tyutyunov, Y., Senina, I., and Arditi, R., 2004. Clustering due to acceleration in the response to population gradient: a simple self-organization model. *American Naturalist*, 164:722–735.

Tyutyunov, Y., Titova, L., and Arditi, R., 2007. A minimal model of pursuit-evasion in a predator-prey system. *Mathematical Modelling of Natural Phenomena*, 2:122–134.

Tyutyunov, Y., Titova, L., and Arditi, R., 2008. Predator interference emerging from trophotaxis in predator-prey systems: an individual-based approach. *Ecological Complexity*, 5:48–58.

Tyutyunov, Y. V., Titova, L. I., Surkov, F. A., and Bakaeva, E. N., 2010. Trophic function of phytophagous rotifers (Rotatoria). Experiment and modelling [in Russian]. *Zhurnal Obshchei Biologii*, 71:52–62.

Tyutyunov, Y. V., Zagrebneva, A. D., Surkov, F. A., and Azovsky, A. I., 2009. Microscale patchiness of the distribution of *Harpacticoida* as a result of trophotaxis. *Biophysics*, 54:355–360.

Tyutyunov, Y. V., Zagrebneva, A. D., Surkov, F. A., and Azovsky, A. I., 2010. Modeling of the population density flow for periodically migrating organisms. *Oceanology*, 50:67–76.

Vander Zanden, M. J., Shuter, B. J., Lester, N., and Rasmussen, J. B., 1999. Patterns of food chain length in lakes: a stable isotope study. *American Naturalist*, 154:406–416.

Veilleux, B. G., 1976. The analysis of a predatory interaction between *Didinium* and *Paramecium*. Master's thesis, University of Alberta.

Veilleux, B. G., 1979. An analysis of the predatory interaction between *Paramecium* and *Didinium*. *Journal of Animal Ecology*, 48:787–803.

Volterra, V., 1931. *Leçons sur la théorie mathématique de la lutte pour la vie*. Paris: Gauthier-Villars.

Vos, M., Kooi, B. W., DeAngelis, D. L., and Mooij, W. M., 2004. Inducible defences and the paradox of enrichment. *Oikos*, 105:471–480.

Vucetich, J. A., Peterson, R. O., and Schaefer, C. L., 2002. The effect of prey and predator densities on wolf predation. *Ecology*, 83:3003–3013.

Walters, C. J., Krause, E., Neill, W. E., and Northcote, T. G., 1987. Equilibrium models for seasonal dynamics of plankton biomass in four oligotrophic lakes. *Canadian Journal of Fisheries and Aquatic Sciences*, 44:1002–1017.

Wardle, D. A., Bardgett, R. D., Callaway, R. M., and Van der Putten, W. H., 2011. Terrestrial ecosystem responses to species gains and losses. *Science*, 332:1273–1277.

Ware, D. M., and Thomson, R. E., 2005. Bottom-up ecosystem trophic dynamics determine fish production in the northeast Pacific. *Science*, 308:1280–1284.

Watson, S., and McCauley, E., 1988. Contrasting patterns of netplankton and nanoplankton production and biomass among lakes. *Canadian Journal of Fisheries and Aquatic Sciences*, 45:915–920.

Watson, S., McCauley, E., and Downing, J. A., 1992. Sigmoid relationships between phosphorus, algal biomass, and algal community structure. *Canadian Journal of Fisheries and Aquatic Sciences*, 49:2605–2610.

Watt, K. E. F., 1959. A mathematical model for the effect of densities of attacked and attacking species on the number attacked. *Canadian Entomologist*, 91:129–144.

Weinberg, S., 1992. *Dreams of a Final Theory*. New York: Pantheon.

Westernhagen, H. v., and Rosenthal, H., 1976. Predator-prey relationship between Pacific herring, *Clupea harengus pallasi*, larvae and a predatory hyperiid amphipod, *Hyperoche medusarum*. *Fishery Bulletin*, 74:669–673.

White, T. R. C., 2005. *Why Does the World Stay Green? Nutrition and Survival of Plant-Eaters*. Melbourne: CSIRO.

Whittaker, R. H., 1975. *Communities and Ecosystems*. New York: Macmillan.

Wiens, J. J., 2011. The causes of species richness patterns across space, time, and clades and the role of "ecological limits". *Quarterly Review of Biology*, 86:75–96.

Wigner, E. P., 1960. The unreasonable effectiveness of mathematics in the natural sciences. *Communications on Pure and Applied Mathematics*, 13:1–14.

Winder, L., Alexander, C. J., Holland, J. M., Woolley, C., and Perry, J. N., 2001. Modelling the dynamic spatio-temporal response of predators to transient prey patches in the field. *Ecology Letters*, 4:568–576.

Yan, N. D., and Strus, R., 1980. Crustacean zooplankton communities of acidic, metal-contaminated lakes near Sudbury, Ontario. *Canadian Journal of Fisheries and Aquatic Sciences*, 37:2282–2293.

INDEX

Abrams, P.A., 4, 23, 26, 61, 80, 115, 127–128, 133
acceleration, 94–97, 101, 105, 107
Adams, J.R., 48
advection, 94–97, 105–106
aggregation (*see also* cluster), 86–87, 94, 104, 107, 113
Akaike information criterion, 47–48, 51, 59
Akçakaya, H.R., 4, 28, 68, 69–70, 73, 115, 146
Akçakaya's lynx model, 111, 125, 137–139
Alces alces, 48
algae, 67, 78–80, 85–87
allometry, 125
Amblyseius degenerans, 43
amplitude, 106, 137
Anarete pritchardi, 95
Anderson, D.R., 51
Anderson, P.W., 126
Anderson, R.M., 107, 116
Andrewartha, H.G., 117
aphid, 36, 43
Aphis craccivora, 43
aquatic ecosystem, 65–68, 78
Arditi, R.
 on biological control, 102, 104, 108
 on enrichment response, 62, 67, 69
 on experimental tests, 85–87, 90
 on instantism, 120
 on model identification, 52, 55, 60, 82
 on predator-prey dynamics, 3, 8, 16, 19, 24
 on trophotaxis, 96, 98–100
 parasitoid attack equation, 39
Arditi-Akçakaya model (AA), 24–25, 28, 40, 87, 111
 emerging from mechanistic model, 98, 100
 fitted to Isle Royale data, 51–55

fitted to literature data, 41–44, 47, 56, 59, 116–117
Arditi-Ginzburg donor control model (AG-DC), 23, 25, 37, 48, 51–52
Arditi-Ginzburg model (AG), 13, 17–18, 20, 23, 25, 27, 40, 111
 emerging from mechanistic model, 98, 100
 fitted to Isle Royale data, 49–55, 122
 fitted to literature data, 56, 59, 61, 117
Arnold, V.I., 117
Arruda, J.A., 72, 73
Artemia, 57
Arvicola, 121
astronomy, 138
attack rate (*see also* searching efficiency), 15
autotaxis. *See* movement and clustering
axiom, 78, 132, 139–140

Bacher, S., 45
bacteria, 9–13, 26, 75, 83
Balanus balanoides, 43
Baltic Sea, 57
bark beetle, 56
barnacle, 38, 43
Batzli, G.O., 121, 138
Baumgaertner, J.U., 19, 25
Beddington, J.R., 3, 29, 102, 104
Begon, M., 4, 115, 117
Bender, E.A., 63–64, 76
benthos, 59, 64, 67
Berezovskaya, F., 20–21
Berger, Leslie Ralph, 13
Bernstein, C., 55
Berryman, A.A., 30, 102, 104, 131
bias, 38–40, 44, 57
biological computer, 118, 122

biological control, 102–107
biological invasion, 140
biomass conversion, 11, 14, 29–31, 121, 146
Birch, L.C., 117
bird, 76, 111
Blaine, T.W., 102
bloom, 78
body size, 125
Bohannan, B.J.M., 9, 74
bootstrap, 47, 51–52, 54
Borer, E.T., 64
bottom-up, 63
Briand, F., 71
Bridgman, Percy, 127
Bulmer, M.G., 82
Burnham, K.P., 51
Busenberg, S., 116

Cabrera, F.M., 74
Calder, W.A., 125–126
Canis lupus, 48
Cantrell, R.S., 102
Carassius auratus, 43
carnivore, 65, 71–72
Carpenter, S.R., 68
carrying capacity, 10, 16, 92, 148
    and biological control, 102–103, 107
    increase of, 22, 77–78, 117
Cartwright, Nancy, 133
Cassida rubiginosa, 45
Caswell, H., 64
Chant, D.A., 43
chaos, 98, 102, 107
Chase, J.M., 68
chemostat, 11, 12, 79, 131
Chesson, P.L., 104
Chlorella, 86
chlorophyll, 70
chrysomelid, 45
cladoceran, 43, 85–86, 101
clerid beetle, 56
climate change, 140
clone, 86
Clupea harangus, 43
cluster (*see also* aggregation), 84–87, 90, 94, 101–102, 105–108
Coccinella septempuncta, 36
coevolution, 80
coexistence, stable, 20–21, 28, 78, 81, 93, 106

Cohen, J.E., 64, 71
Colyvan, M., 6, 82, 117, 120–121, 125–126, 129–131, 134, 139–140
Comins, H.N., 102
competition, 55, 111, 119
concavity, 19, 22, 65, 143–144, 146–147
Contois model, 13, 25–26
Contois, David Ely, 9, 12, 83
convexity, 144, 147
Cook, K., 116
Cooley, T.F., 132
Cosner, C., 101–102
crop, 102–103
crustacean, 43
cycling populations
    and single-species view, 82, 124–126, 149
    in Akçakaya's lynx model, 111, 125, 137
    in Gause experiments, 118

Daphnia, 29, 43, 57, 67, 86, 89
Darwin, Charles, 127
Davies, R.B., 43
Davison, A.C., 51
De Roos, A.M., 106
De Ruiter, P.C., 64
DeAngelis, D.L., 3, 68, 73, 102, 104
DeAngelis-Beddington model (DAB), 25, 51–52, 59, 87, 102
Deevey, E.S., 67, 69
Delia radicum, 43
DeLong, J., 45
Dendroctonus frontalis, 56
depletion, prey, 50, 57, 93
Descartes, René, 127
destabilization, 78–79
detritivores, 69
Didinium, 60, 78, 118, 137
differential equation, 28, 109, 118, 120–124
diffusion, 94–95, 97, 101, 105
disease, 116
donor control, 22–23, 66, 80–82, 89–93, 126

Earth, 133
economics, 126, 132
Edelstein-Keshet, L., 94
edibility, 79–80
Edwards, R.L., 43
Efron, B., 51
Ellner, S.P., 52, 60, 79

Elmhagen, B., 75
Elton, Charles, 124
emergence, 55, 83, 108, 122, 124, 126–127, 131
  of gradual interference, 108–110, 112, 114
  of ratio dependence, 90–93, 101, 108
enrichment response, 62, 66, 71–75, 77, 117, 149
enzyme, 11, 12
Ephestia kühniella, 57
epicycle, 138
epidemiology, 116
Estes, J.A., 66
Eulerian model, 94
eutrophication, 78
Eveleigh, E.S., 43
evidence, standards of, 129
evolutionary elimination, 75, 81, 117–118
exploratory model, 90
exponential growth, 10, 11, 13, 15, 117
  and scale invariance, 130–132
  as fundamental law, 121, 124, 134, 140, 148
extinction, 20, 28, 77–78, 117–119, 125
  and biological control, 103–104
  local, 90, 93, 107, 137

Farmelo, G., 129
Feynman, Richard, 132
fish, 67, 70, 95, 146–147
Fitzpatrick, R., 138
flatworm, 59
Flierl, G., 95, 102
floater, 111
flour beetle, 43
food chain, 62–66, 76, 143–146, 149
  four-level, 67, 71–74, 77, 143
food web, 25, 48, 64, 80–81, 140
forest, 67, 111
fox, 75
Free, C.A., 104
Fretwell, S.D., 71
Fussmann, G.F., 79, 122

Gatto, M., 69, 80
Gause, G.F., 21, 60, 78, 90, 118–120, 125, 137
Genkai-Kato, M., 78, 80
Getz, W.M., 19, 121

Gilliam, J.F., 44–45, 117
Ginzburg, L.R.
  on biomass conversion, 30
  on cyclic dynamics, 82, 125–126, 138
  on enrichment response, 62, 67–70, 72–73
  on home ranges, 26–27, 108, 110, 114
  on instantism, 120–121, 124
  on invariances and laws, 6, 129–131, 133–134, 139
  on paradox of enrichment, 77, 80, 117–118
  on predator-prey dynamics, 3, 4, 8, 16, 19, 24, 61, 115, 128
Gleeson, S.K., 26, 69
Gobler, C.J., 64
Godon, J.J., 13
goldfish, 43
Gompertz, Benjamin, 139
Gotelli, N.J., 4, 115
gradient, 94–96, 105, 107, 112, 114
gradual interference, 25–28, 45, 108, 111, 113–114, 136, 148
grassland, 67
green world, 76
Greene, M., 127
Grünbaum, D., 94
Gurney, W.S.C., 25, 87
Gutierrez, A.P., 19, 25

Hairston, N.G., 63, 76
half-saturation, 11, 12
handling time, 15, 37–44, 47, 107, 143
Hanski, I., 5, 138
Hanson, J.M., 67–70
Hansson, S., 57
hare, 18, 26, 75, 111, 125, 137–138
Harmand, J., 13
Harrison, G.W., 79
Hassell, M.P., 3, 35–36, 41, 43, 55, 61, 104, 107
Hassell-Varley model (HV), 24–25, 36–37, 40–44, 47, 51–52, 87, 116
Hastings, A., 104
Hauzy, C., 102
Hawkins, B.A., 104
herbivore, 33, 46, 64–65, 67, 71, 73, 74, 76
herring, 43

Index [165]

heterogeneity
  and Contois model, 13
  and its stabilizing role, 80
  and population movements, 94, 96, 98–102, 105–108
  as general cause of predator dependence, 18, 83, 87, 89, 93
Hinkley, D.V., 51
Holder, A., 13
Holling model, 15, 18, 25, 27, 40, 83, 90, 105–106
  fitted to Isle Royale data, 51–52, 122
  fitted to literature data, 44, 59
Holling, C.S., 3, 11, 13, 17, 34, 61
Holt, R.D., 9, 65, 116, 131
Holyoak, M., 119
home range, 26, 108–111
homogeneity, required for mass action law, 83, 85–87
Hone, J., 138–139
Hoyer, M.V., 67, 69–70
HSS theory, 76, 77
Hutchinson, G.E., 10, 120
hyena, 23
Hyperoche medusarum, 43

identification, 60, 116–117
implicit equation, 37, 50–56, 116
Inchausti, P., 82, 125
individual-based model, 102, 108, 111
initial conditions, 15, 117
insect, 35, 76, 95, 102, 105
instantaneous form (of the functional response), 35, 50, 57, 59
instantism, 17, 120–124
integrated form (of the functional response), 50, 56–57, 59
interference (definition), 24, 28, 35
invariance, 130–135
Isle Royale, 48–52, 122
isocline
  and biological control paradox, 103–104
  in gradual interference model, 26–28, 108–110
  in Leslie's model, 30–31
  in ratio-dependent model, 20–21
  in Rosenzweig-MacArthur model, 16–17, 118
isopleth, 113–114
isotope, 69
Ivlev, V.S., 25

Jacobian, 80
Jansen, V.A.A., 78, 106
Janson, C.H., 128
Jensen, C.X.J., 4, 14, 26–27, 72–73, 77, 80, 108, 110, 114–115, 118, 122, 124, 133, 138
Jones, J.R., 67, 69–70
Jones, T.H., 41, 43
Jost, C., 13, 20–21, 49–52, 60, 79, 82, 123

Kalff, J., 67–69
Katz, C.H., 38, 42–43
Kaunzinger, C.M.K., 9, 74, 119
Kerfoot, W.C., 68, 73
Kfir, R., 43
King, A.A., 138
Kingsland, S.E., 4, 118–119
Kishida, O., 80
Kolmogorov, Andrey, 3, 135–137
Kratina, P., 59
Krebs, C.J., 4, 116, 139
Kretzschmar, M., 80
Krukonis, G., 126
kudu, 33
Kumar, A., 43

Lacan, I., 57–58
Lagrangian model, 94
lake, 60, 64, 67–74
Lake Superior, 48
Laughlin, R., 126
law, scientific, 126, 132–133, 139
Leggett, W.C., 67, 69
Leibold, M.A., 79–80
Lenski, R.E., 9, 74
Leslie, P.H., 3, 13, 23, 30–31
Levin, S.A., 102
likelihood-ratio test, 47–48
limit cycle
  and biological control, 104, 106–107
  in Akçakaya's lynx model, 125
  in gradual interference dynamic model, 28
  in ratio-dependent dynamic model, 20–21
  in Rosenzweig-MacArthur model, 16–17, 77
limit myth, 139–140, 148
limiting factor, 70, 117, 130, 134, 140, 148
literalism, 123, 124
litterfall, 69

Lobry, C., 13
locust, 94
logistic growth, 10–11, 15, 20, 31, 90, 97, 148
Lotka, Alfred, 3–4, 120, 139
Lotka-Volterra model (definition), 14–15, 25, 52
Luck, R.F., 104
Luckinbill, L.S., 78–79, 90, 119
Lynch, L.D., 104
lynx, 18, 21, 26, 75, 111, 125, 137–138

MacArthur, R.H., 3, 16, 120
macrophyte, 86
Malthus law, 124, 130–131, 140
Malthusian invariance, 134–135
marine environment, 38, 64
mass action, 15, 17, 35, 83, 94, 97, 108
maternal effect, 125–126, 149
May, R.M., 30, 104, 116–117, 129, 135
Maynard Smith, J., 129
Mazumder, A., 79
McCauley, E., 67–69, 79–80
McNaughton, S.J., 67
McQueen, D.J., 68
mechanism
 for Contois functional response, 13
 for donor control, 81, 93
 for elimination of paradox of enrichment, 79
 for gradual interference, 108, 111
 for lynx cycle, 125
 in wolf-moose interaction, 55
 mass action, 15
 maternal effect, 126
 taxis, 94–95, 103
 vs. phenomenon, 126–127
Mertz, D.B., 43
meta-analysis, 45, 64
metaphor, 124, 131
Michaelis-Menten model, 11, 15
Michalski, J., 102
microbiology, 9–14
microcosm, 59, 74–75, 85, 93
midge, 94–96
migration, 98, 101
Mills, E.L., 73–74
Mills, N.J., 57–58
mite, 43, 55, 60
model selection, 48, 50–52
Monod model, 11, 13, 15, 25, 83

Monod, Jacques, 9–12, 26, 149
Monte-Carlo simulation, 113
moon, 133
moose, 48–54, 122–123
Morin, P.J., 9, 74, 119
Mougi, A., 80
Mounier, J., 9
movement
 and biological control, 103, 105–107
 and clustering, 93–101
 and exponential growth, 131
 and gradual interference, 108, 111–114
Murdoch, W.W., 8–9, 79–80, 103–104, 127
Mysis mixta, 57

natural enemy, 102, 103, 105
Nelson, M.I., 13
nested models, 47
Newton law, 131
Newton, Isaac, 127, 138
Nicholson-Bailey model, 35, 37, 39–40, 42
Nisbet, R.M., 80
nitrogen, 67
nonlinear regression, 37, 41, 44, 47, 50, 58–59
Novick, A., 12
null model, 18, 25, 61, 114–115, 128, 139–140
numerical response, 29–30, 32, 91
nutrient
 in bacterial growth, 10–13
 in food chains, 67–70, 73, 75, 79, 146

obstacle, 102
odor, 112, 114
Oksanen, L., 71–75
Okubo, A., 94–96, 105
omnivory, 64
oscillation. *See also* limit cycle
 and maternal effect, 126
 in space, 98–100, 103, 106
outbreak, 104, 106
overcompensation, 24, 36, 40, 47, 52, 54
overfitting, 52, 80, 138
Owen-Smith, N., 33
oxygen, 78

Pace, M.L., 67, 69, 70
paradox of enrichment, 17, 77–80, 117–119
Paramecium, 59, 60, 78, 118, 137
parasitoid, 39–41, 43, 57–59

Parrish, J.K., 94
parsimony, 48, 87, 111, 138
parthenogenesis, 86
partial differential equation, 102, 105, 111
Pascual, M., 102
patch, 90–92
pathogen, 116
Pearl, R., 119
period of cycles, 119, 125–126, 137
periphyton, 102
Perrin, N., 85
persistence, 78, 107
Persson, L., 67, 79
perturbation theory, 90–91
pest insect, 102–107
Peters, R.H., 67–70
Peterson, R.O., 48, 52
Petrovskii, S., 78
Phillips, O.M., 79
phosphorus, 67, 70, 74
Phthorimaea operculella, 43
physics, 121–122, 126, 131–134, 139–140
phytoplankton, 67–68, 70, 73, 146–147
Phytoseiulus persimilis, 36
Pimm, S.L., 22–23, 25, 80, 127
planet, 138
Poggiale, J.C., 89, 91–92
Polis, G.A., 64
Polistes dominulus, 45
pond, 72
Ponsard, S., 55, 63, 69
porcelain, 140
Post, D.M., 71
predator dependence (definition), 8–9, 19, 24
predator removal, 66, 82
Prepas, E.E., 67, 69
prey dependence (definition), 4–5, 8–9, 16, 19
Prigogine, I., 126
primary production
  and food chain length, 71–74
  and Kudu demography, 33
  empirical effect in food chains, 67–68, 75
  theoretical effect in food chains, 62–63, 76–77, 143–144
protozoa, 60, 119
Ptolemy, 138

Quine, W.V., 139

ratio dependence (definition), 3–5, 8–9, 12–13, 17–19
Reeve, J.D., 56
refuge
  dynamic emergence of refuges, 94, 98, 101, 107
  experimental effect, 87
  in Akçakaya's lynx model, 125, 137–138
  stabilizing role, 80, 104
  theoretical effect, 89–93, 102
repulsion, 95
Ricklefs, R.E., 67
Rogers, D., 37, 39
Rosenthal, H., 43
Rosenzweig, M.L., 3, 16–17, 77–78, 117–118, 120
Rosenzweig-MacArthur model, 14
  and biological control, 103, 106–107
  isocline pattern, 16
  limit cycle, 21, 125
  paradox of enrichment, 77, 80
Roth, J., 80
rotifer, 79, 122
Royama, T., 37
rubber, 140
Ruxton, G.D., 25, 87

Saïah, H., 85–86, 90
Sapoukhina, N.Y., 103, 105–107
satiation. *See* saturation
saturation
  absence in donor control, 22–23, 82, 92–93
  and top predator removal, 66
  and underestimation of m, 37–38, 41, 57, 59, 116
  in wasp analysis, 47
  in wolf analysis, 50–52
scaling invariance, 130, 134, 136, 148
Schaffer, W.M., 126, 138
Schenk, D., 45–46, 48
Schiavone, A., 73–74
Seabloom, E.W., 64
searching efficiency
  and biological control, 107
  and gradual interference, 45
  definition, 15
  experimental estimation, 35–42, 57–59
  role in paradox of enrichment, 16, 78

self-organization, 94–95
sharing of prey, 18, 24, 26, 139
   and home ranges, 109–111
shield beetle, 45–46
Shortreed, K.S., 67–69
shrimp, 57
Shurin, J.B., 64–65
sigmoid growth, 10–11
Simocephalus, 86–89
Sitotroga, 43
Skalski, G.T., 44–45, 117
Slobodkin, L.B., 63, 76
Smith, F.E., 63, 76
snail, 38, 43, 102
social structure, 49, 55
solar system, 117
Sommer, F., 64
Sommer, U., 64
Spataro, T., 46–47
species losses, 66, 78, 140
spontaneous generation, 31
stability, 149
   and biological control, 103–107
   and evolutionary elimination, 117–118
   and interference, 24, 28, 109
   in donor-controled systems, 23–24, 80–81, 93
   in prey-dependent systems, 16–17, 77, 79, 125
   in ratio-dependent systems, 20–22
standard model (main properties), 14–15, 21, 62, 77
Stenostomum virginianum, 59
Stibor, H., 64
Stockner, I.G., 67–69
structural stability, 117–118, 130
Strus, R., 67, 69
sun, 117, 133, 138
Sutherland, W.J., 55
swarm, 94–96
symmetry, 130, 132, 134
Szilard, L., 12

taxis. *See* movement
terrestrial ecosystem, 64–68
territoriality, 111
Tetranychus pacificus, 43
Thanasimus dubius, 56
thistle, 46
Thomson, R.E., 67
threshold, pest economic, 103–104

threshold, prey, 26
threshold, repulsion, 95
Thurber, J.M., 52
Tibshirani, R.J., 51
Tilman, D., 81
time delay, time lag, 60, 137
time scale
   and home range, 109–110
   and instantism, 17–18, 120–124
   model with fast and slow scales, 89–92
   of experiments, 33, 34, 61, 82, 84
time series, 21, 60, 82, 122
top level
   and HSS theory, 76–77
   effect of removal, 63–66, 146
   response to primary production, 62, 71, 75, 144
top-down control, 17, 63, 65, 75
Tran, J.K., 27, 133
translation invariance, 132–133
Trew, D.O., 67, 69
triangularity, 80–81
Tribolium castaneum, 43
Trichogramma, 43, 57–58
Trioxys indicus, 43
Tripathi, C.P.M., 43
trophic cascade, 63–66, 75–76, 146, 149
trophotaxis. *See* movement
Trybliographa rapae, 43
Turchin, P., 9, 30, 94, 121, 131, 138
Turnbull, A.L., 43
type-III functional response, 45–48, 59
Tyutyunov, Yu.V., 25, 27, 94–97, 101, 105, 111–114, 122, 133

Urosalphinx cinerea, 43

Vander Zanden, M.J., 71
Varley, G.C., 35–36, 104
Vasseur, D., 45
Veilleux, B.G., 60, 78–79, 119
velocity, 94–96, 105–106, 112, 138
Verhulst, Pierre-François, 10
Volterra, Vito, 3–4, 15, 17, 120, 136, 139
Vos, M., 80
Vucetich, J.A., 48–51

Walters, C.J., 23, 79–80
Wardle, D.A., 140
Ware, D.M., 67

wasp, predatory, 45–46
Watson, S., 79
Watt, K.E.F., 25
Wegener, Alfred, 127
Weinberg, Steven, 140
Westernhagen, H.v., 43
White, T.R.C., 77
Whittaker, R.H., 67
Wiens, J.J., 141
Wigner, Eugene, 132

Winder, L., 107
Winemiller, K.O., 64
wolf, 48–54, 67, 122–123

xylophage, 56

Yamamura, N., 78, 80
Yan, N.D., 67, 69

zooplankton, 57, 60, 67, 70, 73, 85, 146